RELIGION,
THE UNIVERSE
AND EVOLUTION

RELIGION, THE UNIVERSE AND EVOLUTION

A Down-To-Earth, No Nonsense Reality Check

Malcolm Smith

authorHOUSE®

AuthorHouse™ LLC
1663 Liberty Drive
Bloomington, IN 47403
www.authorhouse.com
Phone: 1-800-839-8640

Published by AuthorHouse   07/08/2013

ISBN: 978-1-4817-6759-0 (sc)
ISBN: 978-1-4817-6758-3 (hc)
ISBN: 978-1-4817-6757-6 (e)

Library of Congress Control Number: 2013911257

My Sincere Appreciation and Thanks to

Fred, who sees the best in all of us and whose enthusiasm and encouragement, pushed me to publish this book

and to

Barbara who, while holding very different views, was a model of friendly tolerance

Contents

PART 1   RELIGION

PART 2   THE UNIVERSE

PART 3   EVOLUTION

PART 4   WHAT ABOUT THE FUTURE?

PART 5   WHERE DOES MAN FIT IN?

INTRODUCTION

I am certainly not the first person to be curious about the kind of existence in which we find ourselves. I have often wondered where we come from and where we are going, if anywhere, after we die. Many religions purport to provide answers, but these in turn seem to present more questions than answers. In any case, life is hectic and it is only since retirement that I have had any real, unrushed time to explore these ideas. What follows is my long deferred attempt to shed some light on the matter.

An immediate difficulty was that, unlike researching a high school term paper, there is no accepted source for unbiased, universally accepted information on religion. It was clear that there would be no brief period of research followed by a clear resolution of all my questions. No encyclopedia had an entry under "The Meaning of Life"!! There would be no other option than to use my own brain and reasoning power. For many, a method of last resort!

After thinking about it for a while, I decided to look at the 'big picture.' That would mean investigating three subjects: religion, the universe and evolution. These three should give me some idea of where we came from and perhaps what might be in store for me in the future. I also hoped that the result might provide some guidance as to how to conduct what remains of my time on earth.

What follows is in no way intended to masquerade as an educational treatise on any of the three topics studied. It does, however, trace the thinking I followed in arriving at my conclusions. I have tried always to think logically, scientifically and with as little emotion as possible. It is my hope that the reader will form his own opinions in a similar objective manner, hopefully unswayed by influences of the past.

Much of the information presented here is culled from TV documentaries, books, magazine articles, the internet and memories from my own scholastic education. The views and beliefs are my own and have resulted from my own thinking, but are in no way claimed as original thought.

The reader may notice a slight inconsistency where dates are mentioned; this is because the work was written over the course of a few years and the text represents my views at that specific time.

Where I refer to the Abrahamic God I have used a capital G; when referring to a generic god I have used a lower case g. I have seen the Bible with and without a capital B, I have chosen to use a capital B.

Throughout, I have tried not to include anything of questionable veracity, but at the same time, I have not attempted to verify every last detail. The facts stated are true to the best of my knowledge. The important thing is that they provide an overall, reasonably accurate background in support of my conclusions, which are to be found throughout the text and summarized in Part 5.

PART 1

RELIGION

RELIGION

BACKGROUND

Born in 1935, the first five years of my life were spent in the city of Hull, a large seaport on the east coast of England and a prime target for the German bombers of World War II. When the air raids became too severe in late 1940, the family moved out of the city to the small seaside town of Hornsea; returning to Hull in 1946 after the war. Until that time, I have no memories of any religious influence on me.

In the following years the family: my parents, my sister and I attended the Queen's Hall, a 2,000 seat Methodist church, in the center of the city, with a post war congregation of some two or three hundred . . . . at least on a pleasant Sunday evening in the summer.

As a schoolboy teenager, I belonged to the church's youth club, cycling some 5 miles each way to play table tennis and other enjoyable activities, none of which demanded any particularly Christian dedication. At around 16 I began to wonder about God and whether he really did answer prayers; there having been no results from any of my attempts at prayer so far. I realized that my prayers, beyond the usual Lord's Prayer, were largely self-centered and thus, I reasoned, may not have deserved to be answered. So I devised prayers designed to benefit others and then if they were answered, I would have some sort of evidence to continue 'believing.' Nothing of any concrete nature came of these efforts, in fact, as often as not the reverse came to pass. Of course this led to the question, does God really exist?

None of the folks in the church's activities: Sunday service, Sunday school, the church brass band or youth club were of any assistance in this; all questions to the minister, deaconess, club leader or parents being met with vague non-specific reassurances. I half accepted this, thinking that maybe all would become clear a little later in life . . . . It did not. In any case there was homework to do, soccer games to play, girls to be sought out and much more immediate and more practical demands on my time. **And anyway, at 16 one expects to live forever, so the point was somewhat moot, and certainly of no urgency.**

During my 6 years at university, living in a residence hall, I lived with many young folks professing a strong Christian faith; but again, no one produced any evidence beyond non-verifiable personal experiences. Other folk's solitary personal experiences never did, and still do not, count for much in my book. As a student mathematician and budding scientist I was, and still am, looking for, if not proof, then at least a smidgeon of hard evidence. "You need to have faith" they said, but that seemed to me to be a cop out. Faith is nothing but a belief without evidence . . . or wishful thinking . . . a tenet that will not hold credibility in any scientific discussion, or any court of law for that matter. It seemed to me that faith is more of a comforting 'teddy bear' to hold onto in the absence of any facts.

After university, life became a lot more down to earth; a living to be made, a job to be held onto, sports cars to be driven, squash to be played, new worlds to be explored and immigration. Then a wife, a divorce, promotions, new jobs, new wife, kids, teenagers, another divorce and finally retirement, photography, model airplanes, chess set collecting, bonsai and, finally, serious reading and thinking.

Now, at 73 (in 2008) and no longer expecting to live forever, I am well aware that the sands of time are running out and I am somewhat more interested in what may or may not happen to me after my allotted time expires and me with it.

Over the years I have been, I think, a pretty normal human being. I've made many mistakes, regretted some behaviors, been proud of others and have tried always to be honest with myself . . . not always an easy task. However, I never found the idea of heaven very attractive . . . there just didn't seem to be anything to get excited about. No one seems to have any idea of what it's like, even among those who have earned it, or think they have, including those who claim a close relationship with God. Likewise the idea of hell was not very convincing either, be it eternal fire and agony or "the torment of being separated from God," neither future seemed consistent with an intelligent, all loving, merciful or even a just God.

Ever since my university years I've owned the book: "The Writings of Bertrand Russell," the English mathematician, philosopher and peace activist. Browsing through it recently (2007) I came across his essay, "Why I am not a Christian," and after reading this I realized that I shared many of his views about religion, God and Christianity. Slowly I realized, with a vague feeling of guilt, that I no longer considered myself a Christian, even though I had long ago questioned most specific beliefs associated with Christianity. Not only that, but in a moment of brutal honesty, I finally realized that I had never really positively believed any of the Christian teachings.

Then in 2008 I was watching one of the political networks on TV and happened to see a short interview with Christopher Hitchens who was promoting his latest book "god *(sic)* is not Great—How Religion Poisons Everything." In that TV interview Hitchens' manner was somewhat aloof and take-it-or-leave-it and he did not endear himself to me. However, what he said, did. By contrast, the somewhat superior and condescending attitude of the other interviewee, a Christian believer . . . no doubt brought in by the network to give 'balance' . . . came across very poorly, spouting the same old unconvincing conventional Christian rhetoric. I determined to buy the book and investigate. It would prove to be one of the best decisions I ever made.

THE LITERATURE

To me the book was a hard slog, in the scholarly sense, with paragraph-long sentences, page-long paragraphs and the constant use of words not in common use, requiring frequent reference to the dictionary. Nevertheless I was impressed by the terrible factual indictment he put forward against religions of all flavors and the sorry history of coercion, cruelty and oppression for which they have been responsible over the centuries. All of it in the name of God or Allah or whatever name they give to their own version of a super being. Also the Bible came in for a rough handling, from questions about its origins and veracity to its teachings and numerous contradictions. So strongly does the author berate just about all aspects of the Bible and Christianity, that although finding myself agreeing with what I read, I decided to look for a more subdued atheistic discussion of religion.

A brief look at Amazon's website under 'atheism' yielded, among many hundreds of others, "The God Delusion" by Richard Dawkins. What could fit the bill better than a book written by an Oxford professor?

This 400 pager turned out to be every bit as vitriolic as Hitchens' and covered even more territory, adding to my growing disillusionment in all things religious. More to the point, both books greatly reinforced my long dormant feeling that religion was not for me. However, by this time I was feeling a little brow beaten with all the rigorous, heavy and somewhat pedantic writing. And so, while still quite excited about finding a whole new world of rational thinking, I felt I needed to read how the faithful were reacting to all this criticism.

Another trip to Amazon produced "The REASON for GOD" by Timothy Keller, a theologian and pastor of Redeemer Presbyterian Church in Manhattan. The book had a warm and fuzzy approach and Pastor Keller is clearly a very sincere man, dedicated to his mission. However, I did not get the 'other side of the coin' that I had been hoping for and was left wondering just what the other side was. It seemed to follow many other guidelines I have heard: 'believe and have faith and you will come to believe and have faith' . . . this, in logic theory, is called a circular argument, one that assumes the premise to prove the premise. Actually I felt a sense of disappointment that 'religion' could not come up with anything better than what seemed to me to be no more than the usual 'How-to-achieve-peace-in-this-world-through-God' manual. I felt justified in continuing my venture into atheism.

Still, I decided to give religion another chance. This time I would seek out a more scholarly theological text by someone a little higher up the church's pecking order. I settled upon "God and the New Atheism – A Critical Response to Dawkins, Harris and Hitchens" by John F. Haught, Senior Fellow in Science and Religion at the Woodstock Theological Center at Georgetown University. This looked more like it: a direct reply to the two authors I had just read. The write-up stated that this book 'single handedly dismantles all the skeptics' reasons for doubting' . . . it did not.

The book turned out to be pretty much what I was looking for, but again I was disappointed. The author is somehow unable to make a convincing case for God. The text contains far too many *non sequiturs* and his arguments fall into the 'criticize the other side' type. He gently belittles the 'new atheists' as he calls them, and states that the students in his introductory theology classes would not be 'taken in' by the new atheists' elementary view of theology. He is arguing, in essence, that his opponents take a rather 'blue collar' approach to theology and more sophistication is needed to appreciate the finer aspects of faith. He also frequently and suddenly lapses into 'theo-speak,' my term for paragraphs like the following:

> "Theology can provide a very good answer to why we can trust our minds. We can trust them because, prior to any process of reasoning or empirical inquiry, each of us, simply by virtue of being or existing, is already encompassed by infinite Being, Meaning, Truth, Goodness, and Beauty. We awaken only slowly and obscurely into this unfathomably deep and liberating environment and are bathed in it all our days. We cannot focus on it, and we may not even notice it at all since, like fish in a river, we are so deeply immersed in it. But we may trust our capacity to search for meaning, truth, goodness, and beauty because these have already beckoned and begun to carry us away. Faith, at bottom, is our gracious and enlivening assent to this momentous invitation." (p50)

If this seems a little obscure, read on:

> "How, then, can we justify our cognitional confidence? Not by looking back scientifically at what our minds evolved from, informative as that may be, but only by looking forward toward the infinite meaning and truth looming elusively on the horizon. Simply by reaching toward the fullness of being, truth, goodness, and beauty, we are already in its grasp. This is the true ground of our cognitional confidence, and faith and trust allow us to be drawn toward that horizon in the first place. As we entrust ourselves to the call of being and truth, our minds are already ennobled by the excellence of the goal toward which they are moved. As our minds are being drawn toward truth, these minds already partake of that for which

they hunger. This is what gives our minds the confidence they
need to search for truth." (p51)

This kind of stuff just leaves me stone cold and a little embarrassed that
one would make such flowery, meaningless gobbledygook statements
without any attempt at what the rest of the world calls common logic. It
in no way provides an adequate response to the atheist position. Haught
is correct though in several points. Many of the atheist's arguments are
aimed at the religious fundamentalists, whose primitive beliefs have
been largely left behind by mainstream religion in the US and Europe.
Attacking such beliefs is like shooting fish in a barrel. Secondly, the
atheists lose some credibility by the sheer vitriolic and 'in your face' tone
of their arguments.

**However, leaving aside the vitriol of the atheists and the theo-speak
of the theologians, in my mind theology loses hands down, having
failed to provide any evidence, abstract or real, for the existence of
God. It does, however, produce some inspiring and beautiful wishful
thinking prose.**

I was now feeling more sure of myself but wanted a more down to earth
treatment of the subject and so back to Amazon again! This time I settled
on "Atheist Universe – The Thinking Person's Answer to Christian
Fundamentalism," by David Mills. Although aimed primarily at the
fundamentalists, this book turned out to be what I was looking for. It
provided a wealth of new views, confirming many of my own and shed
light on many of my questions. I now felt justified in considering myself,
if not an atheist, then at least an ***unbeliever***.

Here, I should clarify for the reader, how I (and my Webster's dictionary)
define and use the following terms in this work:

An ***atheist*** is one who asserts the non-existence of any god.

An ***agnostic*** is one who believes it is impossible to know anything about a possible god.

A ***skeptic*** is one who doubts and is critical of all such doctrines.

An ***unbeliever*** is one who sees no credible evidence for the existence of a god.

From this point I expanded my look into atheism, reading several more well rated books and studying many websites devoted to the subject. Websites however seemed to be a little more brash and 'in your face' than I care for. The internet seems to bring the otherwise silent masses out in force! You don't have to look anyone in the eye to post on the 'net, although there is no doubting the sheer volume of testaments to the 'cause.'

I recognize that several books read and studied over the course of a year or so can hardly be called an exhaustive study, but it will suffice for me to consider this a 'prima facie' case for my conclusions.

THE INVENTION OF GODS AND THE EVOLUTION OF RELIGIONS AND CHURCHES.

Ever since the human brain evolved a consciousness and became aware of its own existence and more importantly, its own mortality, man has been apprehensive about the various mishaps and disasters that frequently occur around him. Also, the inevitability of earthly death always loomed ominously on the horizon. Naturally, man would look for any means possible to protect himself as best he could; he knew nothing of science and had no understanding of why or how these frightening events occurred.

Out of such fears came the idea of an intelligent but often vengeful god creating all this mayhem, perhaps as retribution for acts that displeased him. Thus, the god should be appeased by praise, supplication, sacrifice and, of course, monetary offerings.

The very idea of a god is a pretty big concept, and by 'creating' one, a group of people is essentially handing over control of that sphere of influence to the new god. This has been standard procedure by many religions over the ages and these new deities have been given hundreds of names and assigned an even greater number of powers and personalities. Many gods have been specialists, having jurisdiction over specific domains such as war, peace, wine, fertility and just about any area that seemed to need the regulation or oversight of a 'law enforcing' god. Over the years, history has recorded hundreds of gods that have been created, held sway and eventually faded away.

I am not aware of any instance of a god or religion ever being deliberately deposed or otherwise done away with. The end of a god or religion seems to come about for any number of other reasons: lack of need perhaps, because the threat that triggered its creation no longer exists. Maybe the mysterious phenomenon that caused its creation became understood, and therefore no longer needed divine intervention or oversight. Possibly the god failed to produce the desired results and was superseded by a more promising god. Or perhaps the people just outgrew it. Whatever it may have been, sooner or later, they all faded away. As mankind's understanding of the world around him increased, the need for gods declined.

Many researchers point out that most gods were 'gods of gaps,' that is, gods designed to fill the gaps in the peoples' understanding of nature. These gods were abandoned once the 'gap' was filled with an understanding of the problem. It seems to me that this is still the case

11

today; our God is a 'god of gaps' and the number of gaps is steadily decreasing.

It was quite natural that gods would be treated as having humanlike thoughts and attitudes, and presumably capable of dissention among themselves. From this, in man's eyes, would come wars between the gods. This is not unlike the later and current bloody wars between the various 'one true religions.' Having invented a god and made a full commitment to him, then he must be defended against all others, with bloody results.

From this it is clear, at least to me, that man invented God in man's own image, and not the other way round as Christianity and the Bible would have us believe. This is not really surprising since it is almost impossible to conceive of how a being of superior intelligence would think. So He is ascribed all the attributes of a human . . . . only much, much wiser, of course.

Despite all the problems with erratically performing gods and religions, they nevertheless provided man with some imagined protection from the more worrisome aspects of life and death, and thus they survived and even flourished. It seems that the more solace needed by a man, the more inclined he is to believe in the promises of the religion, and the more perceived solace he receives. The whole matter becomes a self-fulfilling invocation.

This combination of need and perceived answer would be quite likely to foster the growth of any religion, and grow they did: hundreds of them around the globe and through the ages.

With the growth of each religion came sophistication, protocol and all the trappings of organization. A pecking order of authority developed and with it means of identification of authority, ceremonial dress and other accoutrements of higher knowledge and authority. Formalized rituals were developed and all the hocus-pocus we see today to separate the herd from the high priests of the Almighty, Allah, Creator, Divine Father or whatever. An essential part of any religion was the emergence of the symbols of belief and doctrine, such as Holy Water, Incense, bread and wine etc. The mysterious aspect was (and is) kept in the forefront by vague concepts such as the Trinity, Saints, Angels, Grace and the Holy Spirit (sounds better than Ghost), and the virgin birth of Jesus, allegedly fathered by God himself (God being a man).

As the Christian religion expanded, there came the need for written works to record the earlier events and to provide a guide for daily living and behavior etc. Unfortunately, the historical documentation and writings available were quite different in scope and content, and often of a contradictory nature. Christianity was a loosely organized group of many different convictions and with much disagreement about the history and validity of the various writings. In particular there were many widely different gospels. This is not surprising considering that some 300 years had elapsed since many of the events in dispute had occurred.

The Roman Emperor Constantine at the time had a dream, we are told, on the eve of the battle of the Milvian Bridge in which he saw a cross in the sky. He won his battle, attributed the victory to God and promptly converted himself and the Roman Empire to Christianity.

Now there is no driving force greater than a new convert (in any walk of life) and Constantine, being the Emperor, had the authority to get things done. It was quite obvious to him that some organization was badly needed if Christianity was to progress in an orderly manner. Accordingly he established a commission in 325AD, the Council of Nicaea, to provide this much-needed organization and specifically to decide which books were indeed the Word of God and worthy of inclusion in the Bible, and which were not divinely inspired and should be officially declared not worthy of inclusion.

These unworthy books were later, in 382AD, not only excluded from the Bible but actually banned and ordered to be destroyed. Over the last few hundred years or so many of these books . . . victims of the literary cleansing . . . have been discovered. This has caused a good deal of consternation within the upper echelons of the church as to how to treat these new revelations and what level of credibility to give them.

The content of the Bible has been the source of much contention through the years and is still hotly debated to this day.

"A collection of writings of unknown date and authorship, rendered into English from supposed copies of supposed originals, unfortunately lost" was Mangasar Magurditch Mangasarian's suggestion for the flyleaf of the King James Bible.

The upper echelons of the Church gained more and more power and of course eventually came to believe in their total power. Guidelines became rules and the Church showed no timidity when it came to explaining all things to suit its purposes. When written down this became dogma . . . the absolute truth because the Church said so.

The Bible has thus been accepted as the 'Word of God' and is therefore often used to 'prove' a particular point asserted within the Bible. This is known, in mathematical logic, as a circular argument and proves nothing.

The first and absolutely essential objective of any organization be it political party, labor union, homeowners association or the local country club, is that of survival. That invariably means an inflow of money and religious organizations are no exception. So, if an organization is to survive and grow it needs financial support and the larger it grows the more money needed. More organizations fail for lack of funds than any other reason; once the money flow dries up, so does the organization. Religions, then and now, like any other endeavor, must keep their current members contributing and must constantly bring in new members. Thus we have the immediate need for dues, renewals, initiation fees, tithing and, of course, "contributions to the work of God in this church" to keep the operation solvent. Some of the recent so-called 'mega churches' have an excellent record in this area. Just one look at the priests' Lincolns and the magnificent glass palaces would make Jesus proud to see that the money extracted from the faithful was being put to such good use.

Of course, to keep the faithful coming back and contributing, a church or religion must offer a reward in return for the people's continued belief, obedience and donations. This reward is the 'knowledge' that the Lord will provide 'meaning' to the everyday tragedies that occur in spite of all the prayers. He will, we are told, look favorably on one in the 'afterlife.' There is, however, usually a warning that commission of certain sins will result in a fate 'terrible to behold' after death, and which will negate all earlier earned credits. This implied threat serves, generally, to maintain membership and to encourage acceptance and compliance with the church's dogma and contribution requirements. The Catholic Church is probably the most successful organization ever in self-perpetuation policies. It goes to great lengths to keep the faithful in line, although, in recent years, the church's grip on the masses seems to be fraying at the edges.

The whole system of organized religion reminds me of a successful Ponzi scheme. Most Ponzi schemes eventually fail when the promised rewards can no longer be paid from contributions. In this case, no

actual reward is ever paid out by the earth-bound religions . . . **God is left to make the payout.**

SUNDAY SCHOOL

I believe that Christianity would not have survived were it not for Sunday school. It is there that the young are schooled (brain washed would be perhaps a better description) in all the beliefs and dogma of the faith. The child is set back from free thought for many years, and sometimes forever. The church has said: "give me a child until he is seven and he is mine forever."

On growing up the young just naturally follow the beliefs and behavior drilled into them. Thus, the absence of any actual evidence is obscured by all the indoctrination and the pressure to follow the crowd. Peer and authoritative pressure are mighty strong forces. To break free or even to question the religious teachings later in life takes an enormous mental and emotional effort. Many will never even try and remain, unthinking, in the flock; rarely attending church, but vaguely affirming their belief in God; never having had any encouragement or opportunity to think their beliefs out for themselves.

I quote from one brief conversation: "I don't want to discuss this, for I'm afraid it may shake my faith."

This is why I believe that there is a case to be made, as Christopher Hitchens has argued, that Sunday school is borderline child abuse.

PROGRESS AND THE "GOD of GAPS"

Throughout the ages, religions have almost invariably been against all science and advancements in knowledge. Unfortunately for them, they have also been in constant retreat in the face of irrefutable advances in science.

Stephen Hawking, the distinguished astrophysicist, recounts how, at an audience with the Pope a few years back, the Pope graciously acknowledged the scientific work that has led to the acceptance of the big bang theory. He then cautioned the scientists present not to carry the study further since the big bang was the creation of the universe and anything before that was the sole province of God!

It is a sad characteristic of all religions to be against all forms of progress, probably because progress is perceived as threatening the power of the religious leaders.

As a result of this mind blocking, I cannot think of a single discovery or invention of benefit to mankind, ever made by any religion.

Good people and good deeds yes, and other forms of wealth redistribution: but in the matter of real progress . . . nothing. It is no wonder that skeptics have dubbed the Almighty as a 'God of Gaps.' In other words, God's province is that of the unknown gaps in man's knowledge of the universe. Once science sheds light on the matter, there is no longer a need for a god to explain that particular mystery.

In the early days of civilization, science was mostly gaps and religions reigned with competition only from other gods. Nowadays, however, there is little territory left for the gods. The largest 'gap' still remaining unexplained is what preceded or caused the big bang. This is a very large gap, although theories abound about the nature of the universe. The most favored currently is that there was no time (or space) before the big bang. This 'gap' important as it is, has little to no bearing on everyday life.

There is a well-known argument commonly called '**The Argument from First Cause**' which goes something like this: Even if you concede that God created the universe at the big bang, you now are faced with explaining how God came into being, and we are back to the child's embarrassing question: "But Daddy, who made God?" A question that is anything but childish.

It may be a while before science fills that particular 'gap.' But then, that's what they said about all the other gaps.

AFTERLIFE

Transition Effects

We hear much talk of a glimpse of the afterlife during the 'process' of dying. The vast majority of these experiences seem to include a bright white light at the end of a tunnel, where often pre-deceased family members are waiting, presumably to welcome the dying person. There is said to be no pain but a pleasant feeling. Naturally all of these experiences are told by survivors of this brush with death and are quite coherent. However, while the experiences are very similar and very convincing to the person involved, it may be just an indication of the way the brain shuts down rather than evidence of an afterlife. After all, who hasn't had dreams where deceased family members seem to be present?

Further, a similar experience has been reported by pilots training in a centrifuge at NASA. During such training the body is exposed to increasing gravity to the point of unconsciousness caused by the forced draining of blood from the brain. It seems that, as the brain becomes starved of blood, the outer parts, containing our most recent memories, are lost first and the older memories are lost last. There is a tunnel-like vision effect with the periphery turning white. The pilots report it as the same white tunnel vision effect experienced under high G forces in a fighter jet.

Others tell of out-of-body experiences where one can look down on one's ailing body and supposedly see things in the room that could otherwise not be seen. I have been unable to find a rational explanation for these reports, except to state that several researchers have noticed that the patient's memory of the details seems to improve for some time after the event.

Heaven

Along with everybody else, I am unable to conjure up any sensible idea of what heaven might be like. Clearly, it is supposed to be at least pleasant, and can be all the way up to 'eternal joy in the love of the Lord.' Well, it sounds nice and maybe it will be, but wouldn't it get boring after a while? It seems to me that existence like that would be for the brain dead. As Susan Ertz (1894-1985) remarked: "millions long for eternity who don't know what to do on a rainy afternoon." I, for one, cannot imagine a condition of existence that I would be happy to enter for all eternity. Therefore, I reason, there must be millions of other folk who feel the same way. So how can the idea of heaven as a reward for earthly 'good, Christian behavior' be much of an incentive? In fact, it doesn't seem to be!!

Many versions of the afterlife have the rewarded ascending into heaven and being reunited with our earthly loved ones who preceded us in death. Unfortunately, on my death, my brain will die and decay and with it my consciousness. So, if I were to survive death in some way, would my cognitive abilities be restored to the condition just prior to death (probably incoherent, babbling and barely thinking)? Or would I be restored at the point of my maximum brain capacity? Or something else? All of this, of course, is pure nonsense; the idea of an afterlife in heaven seems more like a Teddy Bear created to avoid the admittedly frightening idea of finality. Somewhat like buying some high premium after-life insurance.

Hell

What about hell? Hell, to me, takes its place with heaven as one of the predictions of a man-made religion. Reaching its pinnacle before the days of universal education, the threat of being summarily consigned to the flames was seen as a dreadful punishment for straying off the religion-specified narrow path . . . and sometimes not too far off either.

There are many inconsistencies in the idea of hell as threatened by the Bible, Koran and other religions. It ranges from 'eternal burning in the most terrible fires' to 'being separated from the love of God.' More recently, in the last century or two, the religions seem to have realized that it is somewhat inconsistent to draw a line across humanity and to say all those above go to heaven and all below go to hell. What about all those billions of folks who have never been told the rules of the game? And what about children and the brain impaired? Or those of other religions . . . those not of the 'one true God'? Seems a bit harsh to insist that "ignorance of the law is no defense." The state of "limbo" contrived by the Roman Catholic Church into which all those folks are cast, seems a rather silly idea.

So, in the more educated parts of the world there seems to have been a softening of attitude with more attention paid to promoting the 'good' life and hell has taken a back pew . . . a mild embarrassment to religion. **Yet another step in the retreat before rational thinking**.

Recycled

Many religions speak of being born again into this world as someone new. But a reincarnation such as that is, for me, fraught with danger.

When you consider that I was born into a modestly comfortable, healthy, educated, well-fed and white English family, then the idea of a lottery to determine my next life's status among the seven billion humans on earth, are not odds I like very much. Even if you throw in a performance clause, I'm not sure my earthly behavior has been as worthy as that of a sick, semi-starving old man in one of the 'emerging' nations.

Even if my 'afterlife' depends on how 'good' a Christian I have been according to some kind of a reward or punishment scale, it would still not make any sense because I am currently not aware of any previous life and whether or not I am being punished or rewarded for behavior and beliefs during that previous life. How can this be an incentive when no one I have ever met claims to be enjoying the rewards (or punishment) of an earlier life?

My Belief
I see not one scrap of scientific evidence to suggest that the body, mind, brain or soul (whatever that is) survives death, will be rewarded, punished or recycled in any way whatsoever.

My conclusion, so far, is that when you die that's it . . . you're gone . . . . forever . . . game over. What will it be like? Probably like it was before you were born . . . remember?

PRAYER

Does It Work?
Prayer has been one of the mainstays of all religions throughout history. There is no doubt that it can have a great therapeutic effect for countless thousands of people. Everyone feels better after a prayer session, and goes about daily life with renewed vigor and in a happier frame of mind.

However, I have read that despite several scientific studies, no evidence has ever been produced proving that prayer works in any way other than the above predictable and repeatable therapeutic benefits.

Prayer, however is not the only thing that can produce this pleasant effect . . . I have experienced the same kind of effect many times after a ten-mile run. Many others report the same effect on relief from all kinds of stressful events, both physical and mental. Even events as mundane as leaving the dentist's office, completing a grueling assignment, passing a critical examination or getting away with a warning ticket, all seem to leave one in a similar happier, if temporary, frame of mind. There also seems to be a marked similarity between these feelings and the ones experienced by Catholics after 'Confession.'

On the other hand there are countless instances where the prayers of millions of sincere Christians were ignored and millions died. After the December 25, 2004 tsunami, which killed over 250,000, the radio and television waves and churches throughout the world were filled with exhortations to pray for the victims. I chose to donate to the Red Cross knowing beyond any doubt that it would have a far more beneficial effect than any prayer would have had. **God's track record may have been pretty good at feeding a somewhat hungry multitude with a few fishes and a little bread, but it is abysmal when it comes to mitigating the suffering and carnage of major disasters, notwithstanding worldwide prayer.**

An Insult to God's Intelligence?
When God is asked in prayer to provide a humanitarian service or divine intervention of any kind, it seems to imply that God is either still mulling the situation over, sitting on the fence, asleep at the switch or waiting for us to say "please."

Is He not capable of foreseeing the need long before his earthly children do? Invariably a prayer asks for something for ourselves or for others and thus implies that God can be swayed by our requests, or at least needs to be sure of our sincerity! I thought He was supposed to be all-knowing.

Arrogance

Even when I was quite young I remember being somewhat amused and puzzled when 'grace' was said before a meal: "God has blessed us and provided this food etc . . . ." This seemed to me to be arrogance of the first degree; what on earth made us feel that we are worthy of being chosen for a blessing of any kind? What about the starving millions? Maybe we should instead be praying to God to apportion out the world's food in a more equitable manner.

I once observed a children's birthday party held at a local fast food restaurant. The kids were made to shut up while an adult said grace and reminded them that there were many children in the world who were not so fortunate and were starving. Ten minutes later the festivities had been allowed to degenerate into a full-blown food fight. "Glad we don't have to clean this up" was the chaperone's remark to his co-leader.

Likewise, after the usual locker room pre-game prayers; during the game, a player drops down on one knee, in front of 50,000 fans and gives thanks to the Lord for blessing him with a touchdown!

On reflection, I think that much of this type of behavior is more an indictment of the sorry state of education in the world today than it is of anything else. Or maybe it is an indication of just how thoughtless and irrelevant religion has become, despite the 76% of the population that professes to be Christian.

The 76% Christian Nation

Perhaps a word or two about that 76% is in order here. It is often quoted that "we are a Christian nation" and the 76% is offered up as evidence, being the result of numerous surveys and polls taken among "the people."

Now, as mentioned in the paragraph on Sunday school above, the citizen has been subject to indoctrination in Sunday school, school prayers, church, pre-game, pre-race, pre-dinner invocations etc. etc. all of his life. Thus, I suggest, to affirm allegiance to Christianity is an automatic reaction, requiring and receiving no thought whatsoever. Indeed to state otherwise would be to draw immediate and probably unwelcome attention from one's peers. When asked a question by a pollster, or even filling out an anonymous questionnaire, almost everyone will to do the normal, expected and least controversial thing and check the 'Christian' box.

Under these circumstances, what I find remarkable is the 24% who bravely decline to affirm Christianity. 76% is more a measure of how compliant and herd like we are.

"God Bless America"

"God Bless America" . . . I always feel a little uncomfortable when I hear these words. Why do we need to ask God for his blessing? This country justifiably and forcibly separated itself from a repressive and unfair government. It was founded by men (most of whom were not Christians but deists) who believed that they had the opportunity to build a nation based on well thought out, ethical principles and justice. After a few improvements in the Bill of Rights, they succeeded admirably.

The USA, at the time of its founding, wisely and deliberately separated the church from the state. It is still cognizant of that principle. However, religious factions have always tried to usurp the secular leadership and impose their religion's particular views as the guiding principles of life and government in the US. Religion did not build America; Americans dedicated to a peaceful, prosperous, reasonably tolerant and just nation did. God should not receive the credit for building America . . . nor should God be blamed for its warts.

This country, as a result of its size, natural resources, the hard working ethic of its emigrant people, and the beneficial economic rewards of two world wars, (neither of which it started), is the most wealthy in the world, at least for a while. It is also, by far, the most generous nation ever. Further, everyone in the world who wants to escape from oppression, improve an already happy status or get the best college education for his kids looks to the USA. It is nowhere near perfect, but it has grown and evolved in a manner that still owes much to the ethic of its early citizens.

When I hear the words "God bless America," a small voice says to me "well, what about the rest of mankind . . . what do we wish for them?"

The USA has no need to ask for favors or special dispensation from anyone or any deity. Americans are happy to earn them or go without. Fly the flag by all means, but leave it at that.

A Personal Thought

I never could understand the rationale for prayer; if I were the Almighty, the last thing I would want would be My children groveling around in self-abasement over improprieties or even sins committed because of My imperfect creations; or asking Me to set aside My laws of the universe to satisfy their desires, no matter how altruistic.

It seems to me there is nothing healthy or noble about such self-deprecation. Get off your knees, enough already. He is well aware of how great He is and how feeble we are, right? He should, He made us didn't He? . . . . But if that's what you want to do, I will defend your right to believe whatever you wish.

On a lighter note

On a lighter note, I am reminded of the young man who asked his priest whether it was OK to smoke while praying. The priest advised him that it would show disrespect for God and was forbidden. The young man thought for a while and then asked whether it would be OK to pray while smoking.

I see not one scrap of verifiable evidence to suggest that prayer has ever produced any results other than the predictable therapeutic benefits for the supplicant. Nor should it.

ALTRUISM VERSUS RATIONAL SELF-INTEREST

Jesus clearly advocated altruism as the way to achieve heavenly rewards. There are many instances in the bible where He makes this clear. Perhaps the most well-known is the quote "verily, I say unto you, it is harder for a rich man to enter the kingdom of heaven than it is for a camel to pass through the eye of a needle."

The trouble is I think He was misguided in his promotion of altruism. Even as a youngster, I never felt that altruistic values were particularly good for mankind. This is especially true when the recipients are free-loaders or dead-beats. Even if deserved, the transfer of wealth only results in lower standards of living for all. The recipients are rarely able to use the gift to good effect. In other words, what made them poor in the first place will probably continue to hold them down (whether of their own making or not).

The classic example of this would be the USA's welfare system of the late 20th century. The poor came to rely on the handouts, learning to live within the means handed to them and eventually lost all incentive to seek work and earn income. The payments were increasingly regarded as a right and not as a helping hand. Eventually the tax-paying citizens rebelled, the law was changed and the welfare rolls sharply diminished as the rules provided incentive to seek employment. Interestingly, the current Obama administration has weakened this law by relaxing the requirements, and welfare recipients are now on the rise again.

Jesus seemed to look down on the accumulation of wealth. Notice that He didn't just advise generosity, but to give until one was no longer wealthy. Repeating: "It is harder for a rich man to enter the kingdom of heaven than it is for a camel to pass through the eye of a needle."

Consider three people:

Mother Theresa built an incredible ministry and fine reputation in her selfless devotion to healing the sick and destitute. She gave unceasingly of herself and caused many branches of her ministry to be opened worldwide. Mostly with money raised from donations. She died in 1997 but her work continues.

Bill Gates, as a result of foresight, acquired the rights to a new computer software and adapted it to provide the revolutionary Windows graphical user interface format for personal computers. His company, Microsoft, soon became the largest company in the world and, by a 2008 estimate, Bill Gates was the richest man in the world with a net worth of over $90 billion.

He and his wife, Melinda, established a charitable foundation that has donated many **billions** (yes, billions) of dollars to improving the condition of the world's poorest and sickest peoples.

Warren Buffet, thought to be the second richest man in the world, has recently (2008) greatly increased the Gates foundation with a $40 **billion** gift. Incidentally, Warren Buffet is a Wall Street investor, whom Jesus might have regarded as a 'money changer' to be cast out of the temple!

Now Bill Gates, (not a Christian, by the way), and Warren Buffet, both being very rich, may not be able to squeeze their camels through the eye of a needle but I would say that they have benefited the world more than Mother Teresa has. At the very least, God should make an exception for Bill's and Warren's camels.

Altruism has much in common with communism. They both sound like really nice ideals: 'if only we can all work together' kind of stuff, but totally ignore the (God given?) nature of man. The same is true for any animal . . . try asking Rover if he would like to share his dinner with Fido.

Neither altruism nor communism work in practice.

Ayn Rand, in her giant classic "Atlas Shrugged" and other philosophical novels, espoused the doctrine of **'rational self-interest,'** a term that is pretty much self-explanatory. Like most college students who embraced her idealistic ideas, I was much influenced by her writings; I still am, for that matter.

It seems to me that this philosophy of rational self-interest is now the *de facto* basis of most of the economies of the world's developed countries. It is certainly not perfect, but it seems to be the most workable one devised.

Today, Ayn Rand's ideas often come under a good deal of criticism, and her book "The Virtue of Selfishness" is cited with some disdain, mainly because of a clear misunderstanding of the title by folks who have not read the book. In no way is she praising greed or a self-only attitude. Much more acceptable (or 'politically correct') is the term she herself used frequently . . . 'rational self-interest.'

Suppose your employer stops paying your salary, how long will you continue to work for him? Or, alternately when you are offered a higher paying job with a different company, will you stay where you are? To stop working for the first employer and seek a paying job elsewhere and, in the second case, to move to the better paying job is not greed or selfish; it is 'rational self-interest.'

The desire to improve one's situation is not only commendable; it is a cultural trait that is essential if you are to succeed. This basic principle can be extrapolated to the nation as a whole.

Now, not all will succeed, indeed not all *can* succeed because success is a relative term, success for one automatically implies lack of, or less success for another. Equality of opportunity is a commendable goal, but equality of result (where all get a relatively equal share of the wealth) is not. All men are not equal and some will always get rich and others will always remain poor, attempts by governments to redistribute wealth equally to all is doomed to failure.

Such was the fate of the USSR and more recently Greece, where a process of taxing success, providing excessive benefits for all and rewarding laziness has led to poverty, or at least the hard times we now (2013) see for most of Greece's citizenry.

The 'rational' part comes into play when certain minimum standards are set, or a 'safety net' is provided for those in real economic distress. But that should only become a way of life for those unable to fend for themselves. This is admittedly a difficult goal to manage.

Any nation that has become wealthy, and whose citizens enjoy a high standard of living, became wealthy as a direct result of the industrial production and the commercial efforts of its citizen entrepreneurs and its corporations. As the wealth builds up, initially in the few, the benefits in the form of a higher standard of living slowly but steadily flows to the general population. To force the process is doomed to failure.

What seems to happen is that as a nation builds up wealth under relatively low taxation and minimal regulation, only the few initially benefit. Then there follows a change in government policy, and taxation, regulation and wealth redistribution all increase markedly under the banner that all must share in the prosperity. The result is that the pendulum swings the opposite way and the total wealth production slows down under the new restrictions. Then, in the global free market, the nations that are still in the unfettered wealth-building phase, out-produce the more established nations and even surpass them in economic strength. Witness the rise and decline in economic strength of certain parts of Europe vis-à-vis the emergence of Japan in the 60s and now China and other 'emerging' nations today.

It has proven hard for nations to strike the right balance in governmental control between absolute freedom from regulation and strangulation by taxes and red tape. What usually happens is that the national wealth is built under a regime of minimal regulation. Examples would be the industrial revolution in Britain and the emergence of the railroads and industrial might in the US after the world wars. Now the pendulum is starting to swing the other way, as it has already done in Europe.

On the other hand, countries run by religious dictators, or held to religiously inspired commandments, hardly ever get out of the 7th century way of thinking. Even in modern wealthy nations I cannot think of one instance where the church has contributed to the improvement in the general population's standard of living. The churches are, however, adept at redistributing the wealth produced by others.

Any code of behavior must complement man's nature; it never works the other way round.

TOLERANCE AND TERRORISM

I believe that other people's religious opinions, speech, proselytizing and deeds should be tolerated to the extent that they cause no harm to others or become a frequent irritation. Few would disagree with this statement but, as usual, the devil is in the details.

The Islamic religion has been around for as long as Christianity and currently a great many people in the world live under its influence. It is based on the Koran which has much in common with the Bible, although scholars apparently disagree on much of the actual translation from the original Koran documents, many of which have been lost. Nevertheless the Koran, as promulgated by its fundamentalist faithful, is today very militant and intolerant of other religions. Certain passages of it specifically demand that (Koran) Believers kill Non-(Koran)-Believers, or Infidels . . . tolerance is not an option. I understand that, in the minds of the Clerics, these passages are quite clear and leave no room for interpretation.

A majority of the Islamic peoples treat these passages in the same way that Christian believers treat much of the Bible . . . somewhat like the 800 pound gorilla in the room: yes, it's there but ignored. However, there is a large and growing faction, loosely called the Islamic extremists, that takes this commandment to heart.

For hundreds of years, throughout the Arab world, even back to the roots of Islam in the 7th century numerous factions, tribes and even nations

31

have waged war against each other continuously; all in the name of jihad, or religious war. All too often these wars have been wars over secular differences and Koran commandments have been the excuse to justify such wars.

Whether or not the Koran actually does order such brutality is a matter for the Islam clergy to interpret for the faithful. In the western world, such religious commandments would today receive a lot of 'push-back' and opposition, especially in the democratic nations. In the Arab world, however, things are different.

Education in much of the Islamic world is a far cry from schooling in the west. Girls and women are, to all intents and purposes, second class citizens. In many Arab countries they receive the barest of schooling, are required to be subservient to their menfolk and have virtually no vote or influence on the society. Indeed, wives are considered the property of the husband . . . and treated accordingly.

Boys on the other hand are required to spend enormous amounts of time reading, reciting and memorizing large passages from the Koran and are taught that this is the life they must lead without dissent of any kind. The Koran is not just a religion but a whole, complete way of life, and obedience to the leaders' interpretations of it is paramount. Any straying from this narrow path is dealt with harshly. Other aspects of education, held important by the West, receive sparse attention.

When one considers the results of the childhood indoctrination into the ways of the Christian church in the western world, it is not difficult to imagine the slave-like products of the Islamic boys' schools. When instilled with a hatred for all Infidels, especially the US Infidels, you have what amounts to a terrorist factory.

Given the low standard of living and education in many parts of the Muslim world, coupled with the sight of all the western luxuries of the West on TV, it is no wonder that the young men flock to join the jihad.

The Koran makes the situation worse by promising these militants, in return for their heroic death, an orgy-filled afterlife, and glory to the surviving families of all who die for the 'cause.'

Unfortunately, the terrorist militants are not a national 'army' that could be attacked and defeated by conventional war tactics but are organizations commingling within the civilian population in every country where Islam is a significant religion.

It is my opinion that if the US were to change its approach to a more hands off, but still strongly defensive stance, much of the resentment against us would abate and the terrorism against the West will die down like a fire without fuel. The 'cause' will lose its attraction for the young.

Unfortunately as the US enjoys some success in the fight against terrorism, the terrorists are now spreading out to the less developed and poorer parts of the world, notably in Africa. So we may expect the horrors to continue for some time as these terrorists recruit willing Islamic youth from these poverty stricken parts of the world.

This 'cause' may have been helped considerably by the perceived arrogant and somewhat domineering attitude of the USA in recent years (1990 through the present at least) in its attempts to play world cop and bring 'democracy' to the world.

Universal, world-wide secular education and the promotion of rational free thought for the world's youth is probably the best way to combat religious extremes . . . . One can dream.

DEMOCRACY FOR ALL?

Westerners are brought up and schooled to believe that democracy is the one true and great form of government. A few words discussing this may be appropriate here.

Many scholars and students of government and civilizations have spent countless hours writing on and debating the most desirable form of government. This writer does not, for one minute, profess to be an expert in these matters. However certain pros and cons about democracy appear to be obvious.

The most desirable point in favor of democracy seems to be, not its effectiveness or even its soundness, but its stability. So long as everyone gets a vote and a simple majority wins, the people will never be saddled with tyranny or incompetence for long.

The old saying is still true: *"you can fool all the people some of the time and you can even fool some of the people all the time, but you cannot fool all the people all the time."*

Unfortunately, the ones who potentially would make the best leaders are rarely the ones who emerge through the party system to the level of presidential candidacy. Witness some of the comical candidates of the last 20 years or so. He who promises most has a built-in advantage at the get-go.

Another problem with democracy is that it gives the uneducated, drunken lazy jerk exactly the same say in who runs the country as the well-educated thinking person with the best interests of the community at heart. Well they say, 'it all evens out.' True enough and that is what we get: a 'mediocracy.' How many presidents in the USA have we had the fit the ideal?

A problem that is becoming more serious is that as the parties have diverged and become more and more inflexible, the politics have become more and more a struggle for re-election. So much so that the law of the land is being contested according to how many voters it will add to each party's voting roster. Add to this the third factor of the presidency and you have the recipe for gridlock. This rarely happened at the time of the founders, who came, served, and returned home; there were few real career politicians. But today things are a lot more complicated and with jobs and livelihoods at stake for the leaders themselves, debate becomes a lot more self-centered.

Nevertheless, it is still unlikely that a bad leadership will last for very long. But it may be replaced by one just as bad!

Other nations have managed quite well with kingdoms or oligarchies and even dictatorships. A benign dictatorship where the ruler is committed to the well-being of the people is probably the best government of all. There are two main problems with even a benign dictatorship: the first is when the nation is just too big and complex for one person to govern. The second problem comes with succession, the dictator rarely has the ability to choose someone as competent as he, and nepotism will produce incompetence eventually. North Korea seems to be one current situation like this.

Regardless of all the pros and cons about any form of government, no one nation has the right to tell another how to govern itself. If one nation mistreats its citizens to the extent that another nation feels the need to intrude, it would do well to think out the 'unintended consequences' before acting. Rarely is such intervention welcomed after the initial overthrow. Generally the bad guy is succeeded by another almost as bad, or the US places a favorite (pro—US) in place. Rarely does the intervention result in a lasting successful change.

Perhaps we should adopt a tacit recognition that democracy is not all it is made out to be and allow other nations to govern themselves as they see fit.

MIRACLES

The Bible tells of several miracles performed by Jesus. Those coming to mind are the feeding of the multitudes from a couple of fish and five loaves (or was it the other way round?); Jesus walking on water; the casting out of demons and the curing of the bedridden. All of these seem intended to demonstrate the all-powerful nature of Jesus' ability and at the same time perform some humane service.

When these are cited as testimony to the Almighty's power through Jesus, I always wonder, why a miracle on that particular occasion? None of them are of earth-shattering importance. And why no miracles on other equally or more deserving occasions? It seems to me that most could have become miracles through the retelling many times . . . like modern fish stories. Also, many of the miracle cures were conducted on folks with psychiatric type ailments, the kind which today are cured through counseling, although admittedly over a longer period of time. Why were no miracles performed where lost limbs were restored? Such events would have provided a much more compelling reason to believe.

Many people today attest to miracles they claim to have observed or personally experienced. The problem with all these experiences is that none of them ever leaves any persuasive evidence that others may review. Most seem to recount a narrow escape from some impending injury and usually produces a feeling of the need to thank someone for the result. The lucky individual seizes on this as proof of the Almighty's intervention. I too survived many such occasions but on reflection, I think it is just a normal reaction triggered by relief at having lived through the danger . . . . And, of course, from childhood's parental admonishments to thank God for everything.

DISASTERS

I have noticed that religious minded survivors of almost any tragedy or disaster are always quick to thank their God for sparing them, even though their homes and all their possessions may have been lost. If God was indeed involved and perhaps directed and orchestrated the event, it would seem to be much more rational to regard the result as more of a warning shot across the bow, rather than as a blessing.

At Thanksgiving, the faithful give thanks for the successful harvesting of their crops, God having looked kindly on them. Do they also believe that God was meting out punishment to the folks in the Midwest whose crops were all lost in the drought?

In destroying 250,000 people in the December 26, 2004 tsunami, how did God decide who to save and who should perish?

Millions every year, and for centuries past, all over the world, die horrible, excruciatingly painful deaths, often at the hands of other God's children. We are told that God has a 'higher purpose' in mind while he watches over this carnage. When you include children in the victim list, one cannot help but wonder what 'higher purpose' justifies such horror.

"The love of God passeth all understanding" . . . **it doth indeed.**

If I am wrong in these beliefs and am called upon to confront God on my death, it is God who will have the bulk of the explaining to do.

THE BIBLE AND JESUS

If ever there was a book written intended to drive people away from Christianity, it would have to be the Bible.

In my college days, being curious about God and Christianity, I decided to actually read the bible and find out what it was all about once and for all. It proved to be a mistake. I attacked the project with all the vigor of a twenty year old and waded through it in about three weeks, sacrificing many hours that could have been spent more pleasurably on the squash court. At the end I was no wiser, had no idea what it was all about, and was quite taken aback by the apparent lack of any sense of order or intended message. But . . . I had read it . . . mission accomplished.

Within the beautiful, soft white leather bound, gold cross embossed, volume of gilt edged pages, there is a great deal of beautiful prose and a lot of great advice. There is also the depiction of God as a top rated unmitigated monster.

Like it or not, the bible has some very real authenticity problems. Of course the many translations from scripts that really do not translate too well, coupled with the countless revisions and 'interpretations' (by believers, of course) over two thousand years, have added to the confusion. Not to mention the battles over the canonicity of many of the books, both included and excluded.

It has received much valid criticism, most succinct is perhaps that implied by Mangasar Magurditch Mangasarian's suggestion for the dustcover of the King James Bible:

> *"A collection of writings of unknown date and authorship, rendered into English from supposed copies of supposed originals, unfortunately lost."*

A review of Wikipedia's section on the Bible will yield a gold mine of information both pro and con regarding the Bible. The reader will be left as confused as ever. That, of course is where the clergy comes in, to explain it all.

Today, I am left with a series of impressions, perhaps they are best just listed with minimum comment:

- It has many contradictions, half of which must be errors.
- It hopelessly misses the dates, sequence and manner of the creation of earth.
- Books are omitted and/or canonized by what authority? The same authority that brought us the inquisition? Sounds more like 'literary cleansing' to me.
- There is still an ongoing disagreement over the meaning of virgin (vs. young unmarried woman, virginity optional).
- Jesus abandoned his family.
- Jesus married / mistress, fathered children? All hotly argued today.
- Jesus many times spoke of his immediate or impending return within the lifetime of those present. He never showed up.
- Jesus seems to have had a short fuse.
- Miracles seem random, sometimes for no good reason. Why no real indisputable ones, like a replacement limb or two?
- Jesus died on the cross that we might be saved. **Where is the logic in that? That phrase seems to me to be the non sequitur of all time**. And saved from what . . . hell?
- Maybe he didn't die on the cross?
- Jesus seemed to believe in hell and everlasting punishment.

- Jesus seemed to believe altruism was the way to go.
- None of the original books, written 40+ years after the events, exist today.
- Jesus seemed to believe in (or at least accept) the subjugation of women.
- The Bible clearly supports religious intolerance.
- Capital punishment is acceptable throughout.
- Incest is tolerated in certain parts.
- Slavery is acceptable in both old and new testaments.
- Certain religious wars are obligatory (shades of Islam extremists?)
- God orders certain genocides: Canaanites and Amalekites.
- Some of the miracle predictions were 'predicted' after the event.
- Some prophesies not fulfilled . . . the destruction of Tyre.
- Judaism does not consider Jesus to be qualified to be the Redeemer.

The Bible has one great feature however; it is perfect for the cherry-picking religions like the Jehovah's Witnesses. These folks delight in finding just the right 'scripture' for any occasion, regardless of the context of the original.

REQUIRED CHRISTIAN BELIEFS

I don't recall seeing a specific list of 'beliefs' required to be held by a full-blown, card-carrying, dues paying Christian. The Ten Commandments seem to be just a code of behavior handed out to an ill-educated group of folks in difficult times. In any case, there were no Christians in those days. Still, the following is what comes to mind:

- Acknowledgement of the only one true God.
- Belief that God is all loving, all-powerful, all-knowing and all-merciful.

- Need to worship this God.
- Belief in the virgin birth.
- Belief that Jesus was the son of God.
- Belief in miracles.
- Belief in the crucifixion, the resurrection and the ascension.
- Belief in an afterlife.
- A fervent conviction of the truth and personal relevance of the above.

I cannot claim even one of the above for my resume.

COULD THERE BE A GOD?

Could there be a God? Possibly, but this admission is forced only because it is impossible to prove a negative or the non-existence of God.

Still, if He exists, He has gone to extraordinary lengths to disguise the fact. He has succeeded in preventing even the tiniest fraction of positive evidence to leak out. Why give us a brain if we are supposed to check it at the church door and take everything on faith? Faith is, after all, just belief without evidence.

SOME TROUBLING THOUGHTS

What of the preachers, priests and other men of God? If there is no God, then he does not speak to them. Are they so self-deluded that they really do believe that God is speaking to them? Do they live a life of torment knowing that all the years of preaching are a sham and for naught? Is there no feeling of guilt? Or do they perhaps rest happy in the thought that they give comfort to many . . . even though delusional?

IN CONCLUSION

So, after all this, where do I find myself? To put it simply:

Throughout my life so far, I have never heard, read, seen or experienced the slightest scrap of evidence, not explainable by natural causes, to indicate the existence of a benevolent, malevolent or even disinterested God.

This is not to say that I am convinced that there is no god, I just do not see any scientific evidence in support of the existence of any kind of god . . . . **Nor do I see any reason for one.**

I need to look elsewhere.

PART 2

THE UNIVERSE

THE UNIVERSE

BACKGROUND

When I was 14 or 15 I was taught something about the solar system in school and learned that Galileo had revolutionized astronomy by designing a telescope that would magnify up to about 30x.

The design of the telescope he used was discussed in the physics optics class, and seemed simple enough to me, so I decided to try to make one. A visit to the local optometrist yielded a basic circular lens of one meter focal length and about 2 inches in diameter. Nothing larger was available since they were optometrists producing just spectacles, not scientific instruments. I also needed a smaller diameter lens of much shorter focal length to provide the required magnification. After much thinking, I hit upon the idea of borrowing a lens out of my microscope. Now all I needed was an enclosure. For this, I used two telescoping cardboard tubes with the lenses taped to each end. The whole contraption was about three feet long.

I vividly remember holding it up to my bedroom window, on a clear night, and fiddling with the telescoping tubes. What had always been a spec of light to my naked eye suddenly became a clearly defined circular disc. It had to be Jupiter! My bedroom faced west and the star chart I had said that that was where Jupiter should be. I was thrilled that I could actually see another planet. I even convinced myself that I could see four of Jupiter's moons. From that day on, I was hooked on astronomy and cosmology.

Maybe a year later, my friend and I joined the British Interplanetary Society, a group given over to exploring the possibility of interplanetary travel. Most of the proceedings published were beyond us but we felt part of the future.

I bought a wonderful book entitled "The Nature of the Universe" by Fred Hoyle, a noted Cambridge University astrophysicist. It had a full color cover photo of a total eclipse of the sun surrounded by the fiery corona. I loved that book, reading it over and over. I felt that this book was just about all that could be written about the universe. In it, Hoyle described, in down to earth terms, his theory of continuous creation producing a 'steady state' universe. I remember that he argued that one atom created per year in the volume equivalent to that of St. Paul's cathedral would account for all the mass of the known universe and would explain the observed expansion of the universe. It was a great explanation of how everything came into being. The theory was elegant, neat, tidy, accounted for just about all current observations, and made good sense. Unfortunately, it was also wrong.

Later, Hoyle's theory of continuous creation lost out to the theory of a singularity that suddenly exploded and expanded into the universe. Hoyle rejected this new theory and derisively termed it creation by a 'big bang.' Ironically, this term would stick and is now the accepted name for the origin of the universe. Hoyle, however, never did accept the 'big bang' theory.

THE SCIENTIFIC METHOD

A word or two about the 'scientific method' is in order here. Everybody knows what a theory is; it is a suggested explanation for something that is not understood. It can be the carefully considered result of much study or it can be a wild, unthinking guess. A theory, by itself, does not carry much weight, probably no more than the reputation of the person proposing it. So, through the years scientists have come to agree on certain rules by which a theory will be judged. The procedure is pretty much as follows:

1) The problem is explicitly identified.
2) All available facts are assembled and studied.
3) At this point, a theory is developed to fit all the facts. The theory may indeed be no more than a guess, providing it meets all the facts and does not contravene any previously established laws.
4) Next, experiments are planned and the results are predicted according to the theory.
5) The experiments are then carried out and if the results are consistent with those predicted by the theory, then there is cause for optimism and more experiments are devised to add further confirmation.
6) Any unsuccessful results are cause for abandoning, or at least modifying, the theory.
7) Eventually enough successful evidence may be amassed that the scientist is convinced of the soundness of the theory and he publishes his results and presents his findings for the world to review, reproduce, criticize and acclaim or denounce (with appropriate reason).

Few, if any, theories reach the level of absolute fact. Generally, the best that one can achieve is that of undisputed acceptance. However, all theories are always subject to future discoveries or developments. This is true for much of what is written in these pages.

Science does not lay down a concrete road, but rather fashions what seems like the most promising path. It remains ever ready to backtrack and, if necessary, explore a different path . . . . albeit with much wailing and gnashing of teeth.

THE DOPPLER EFFECT

Before going further, it would be helpful to understand a phenomenon that has played an important part in man's study of the universe.

Stand by any roadside and listen to the sound of cars passing at high speed. The tone of the sound takes a sharp drop as the car goes by. Consider the sound emitted by the car when it is still, say, 100 yards from you. By the time the sound reaches you, the car will already have moved a small distance closer to you. This means that the sound waves emitted during this time will be slightly squashed up or compressed and will have a slightly shorter wavelength and therefore a slightly higher pitch. After the car passes, the reverse is true and the sound waves are stretched out and have a lower pitch. Hence, the sudden drop in pitch of the sound. This is called the Doppler Effect. The same thing happens to light waves but at a much higher speed of course.

Now white light is just a mixture of all the colors of the rainbow, and in astronomy, the light is sometimes broken down into its color components by using a prism, like the one you played with as a kid. Red light is of lower frequency and blue light is of higher frequency.

Let's take a photograph through a prism in a telescope of a distant galaxy. Now let's do the same for our own galaxy, the Milky Way. When we compare the two photos we'll see a strange effect. The distant galaxy will look a lot more red than ours. The light from the far galaxy has been stretched out meaning that the galaxy is moving away from us. By measuring the amount of this so-called Doppler 'red shift,' we can figure out how fast the far galaxy is receding from ours.

LOOKING BACK TO THE BEGINNING

Astronomers have found that not only are all the galaxies receding from us, but they are also all receding from each other, and the further away they are from each other, the faster they are receding.

Also the further away they are means that light has taken a lot longer to get here and we are actually looking at a distant galaxy the way it was when the light left it. In other words, we are looking back in time. And many millions of years ago. So, by looking at the more distant galaxies, we are looking further and further back in time. The photos of distant objects show them at a much younger age and show us a much less developed universe than we have closer to us today.

By doing extensive study and measurements, scientists were able to formulate mathematical laws governing the behavior of the universe in its younger days. Now, if everything is moving away from everything else; then in the past, everything must have been closer to everything else, and so, by working backwards, it was deduced that everything came together at one specific time in the past.

Beyond a certain point, however, observation is no longer possible with our current technology, and we must use advanced mathematics to predict what happened in the early stages of the universe. So sophisticated are these techniques that scientists have formulated a depiction of the universe all the way back to when the universe was just 10^{-43} of a second old. At this point the temperature of the universe was 10^{32} degrees Kelvin; and its density was 10^{96} times that of water . . . . hmm.

THE ORIGIN OF THE UNIVERSE

If you are devoted to the literal truth of the Bible, you believe you already know how the universe came into being. Genesis does a great job in describing, at least in somewhat general terms, the events of the first seven days. You may also be aware that Bible scholars, Archbishop Ussher in 1650 among others, have used the information in the Bible to determine the actual date of creation. We are told that it occurred in 4004BC on October 23rd . . . . in the afternoon, although the actual time was not disclosed.

Not bad for folks trying to account for a creation that occurred some 60 'begats' ago.

On the other hand if you prefer a more scientific approach:

Around 13.7 billion years ago, the universe came into being from what is called (for lack of a more precise or understandable term) a "singularity." Now a singularity is a term used by mathematicians and scientists to describe a point where certain values become infinite or indeterminate, meaning that the equations and laws fail, are no longer valid, and cannot predict events beyond that point . . . . In other words, they don't really know what happened.

In the case of the universe, the singularity is described as an infinitesimally small, infinitely dense point that suddenly exploded into existence. This event has been called the "big bang." It is asserted by some, that both time and space was initiated at this time. It is also asserted (by the same some) that since space and time both came into existence with the big bang, it is meaningless to speak of a time before then . . . . hmm.

However, while most scientists agree that tracing the universe back to a tiny fraction of time after the big bang seems to be supported by observations and is generally accepted, there is disagreement as to whether time itself started at that instant. Currently there is a fashionably modern school of theoretical physicists that advocates the "string theory." This theory suggests that there may be a multitude of parallel universes existing in the form of multi-dimensional membranes or "branes," and that it was the collision of two such entities that triggered the big bang. Thus, they argue, while the universe as we know it may have started at the big bang, time itself may have preceded that event.

This stuff is all very esoteric and largely (if not wholly), mathematics based. Thus, while interesting, it does not really belong here in this review of what we actually know of the universe. More knowledge or tools will be needed to discover what, if anything, happened to cause the big bang.

I am not willing to fall into the same trap that has befallen hundreds of prior civilizations by assigning everything we do not understand to a god or other supernatural being; only to find a perfectly natural explanation later. Almost all religions, particularly the Abrahamic faiths, have been in full and ignominious retreat before scientific knowledge since day one. Even as recently as 1981, the Pope acknowledged that it was OK for science to pursue knowledge back to the big bang, but that was God's creation and should not be examined further! (*Stephen Hawking in "A Brief History of Time"*). This is an example of the so-called "God of Gaps" . . . any gap in our knowledge is routinely assigned to a god.

Nevertheless, there are those who argue, like the Pope, that this, in itself, is evidence of the work of God or other supernatural cause of the universe. But now, at the frontier of knowledge, I have no intention of invoking the 'God of Gaps.' I will just wait a few years for more information.

THE EARLY DEVELOPMENT OF THE UNIVERSE

<u>At the instant</u> of the big bang the universe was infinitesimally small, infinitely dense and infinitely hot. Space-time (a single entity) was infinitely curved and occupied no space or time. Pretty small.

<u>After one second,</u> the temperature had dropped to about 10 billion degrees Celsius and photons, electrons, neutrinos, protons and neutrons had all made an appearance. These are much more fundamental particles than atoms or molecules and are needed before more complex particles can form.

<u>After 100 seconds</u> the universe had cooled to a somewhat chilly 1 billion degrees. Heavy hydrogen (deuterium), helium and some lithium and beryllium had formed.

During this time the baby universe emitted vast amounts of microwave radiation, which spread out and cooled down. The remnants of this radiation still exist and have been detected, measured and mapped. Today its temperature is only a few degrees above the absolute zero. It also exhibits minor temperature variations and is not perfectly uniform.

THE BIRTH OF STARS AND GALAXIES

During the next million years or so, the universe continued its expansion and cooling. Its form was that of a nearly homogeneous, massive, expanding cloud. As the expansion continued, so did the cooling and the drop in energy level. Eventually, atoms started to form and the energy reduction due to expansion allowed gravity to have an increasing effect and produce locally more dense areas. These denser areas would have a greater internal gravitational effect and would tend to separate into large clouds and eventually condense into stars and galaxies.

Stars and galaxies developed simultaneously and their formation continues to this day.

As the clouds of matter formed the beginnings of a star or galaxy, a rotation would develop. This rotation would balance the effect of the force of gravity, which would grow stronger as the cloud closed in on its center. All stars and galaxies have rotation, which tries to throw the matter outwards and so counteract the inwardly acting gravitational effect. Eventually a balance and stability is attained, which we see in the many galaxies, stars, planetary and moon systems throughout the universe.

As the locally denser areas of the universe (consisting mostly of hydrogen and helium) separated out and started to condense under the force of gravity, the temperature and internal pressure of the cloud so formed would rise. Eventually the pressure and temperature became so great that the hydrogen atoms began nuclear fusion, producing helium with a corresponding release of enormous amounts of energy, (this is what happens in a hydrogen bomb). Again, the enormous expansive effect of all this released energy becomes balanced by the gravitational forces trying to collapse it. The situation becomes stable and **a star is born**.

The quantity of gas captured in this condensing process determines the size of the star and also the intensity of the nuclear reaction. Just like a bonfire, the more wood you put on, the more intensely it burns but the sooner you run out of fuel and the fire goes out. Likewise, the larger the star, the shorter its life. Our star, the sun, is a medium sized star, about 5 billion years old and is about half way through its life. Another 5 billion years and it will run out of gas and die. In the meantime, we should be looking for somewhere else to live.

Our sun is probably a third generation star in that the gas cloud from which it formed, itself contained the remnants of a previous supernova explosion (see below).

A VARIETY OF STARS

There are many kinds of stars, depending mostly on each's mass, and each follows a somewhat different life cycle, here are but a few:

<u>Stars smaller than the sun</u> are believed to remain basically unchanged until they run out of fuel and just burn out to a lifeless mass. They are known as "**brown dwarfs**,"

<u>Stars of approximately the mass of our sun (one solar mass)</u> steadily burn their hydrogen, forming a helium core. As the outer shell of burning hydrogen expands, the core turns to helium. The resulting expansion results in what is called a "**red giant**." In our case, the sun will swell to just about swallow the earth; incineration is certain, at least. Eventually our sun and others like it will shrink down and develop a carbon core. It will temporarily gain brightness as it reheats under the gravitational pressure. At this stage it will be known as a "**white dwarf**."

<u>Stars of 10 to 30 solar masses</u> also form a helium core within a burning gaseous shell. These stars also expand, but to a size far greater than the red giants; they are called **super giants**. As they run out of fuel they collapse in on themselves to form, small, very dense, **neutron stars**.

<u>Stars of greater than 30 solar masses</u> follow the same route except that the extreme intensity of their cores results in the production of all the heavier elements. As the star runs out of fuel, its inward collapse results in an enormous explosion called a **supernova.** It is this final intense

explosion that scatters into space all the other heavier elements, dust and debris found throughout the universe.

Although common in the universe, it is rare that a supernova is actually seen from earth. However, there are old Chinese art works depicting a bright event in the sky in the 11th century. Also, in the abandoned caves of the Anastasi people, there are wall paintings depicting a very bright 'star' that shone briefly in the heavens. These paintings have been carbon dated to 1054AD when, on July 4th, there was a supernova explosion in the Crab constellation. It is the remnants of this supernova that is now called the Crab nebula. The gas from this explosion is still expanding at the rate of 3 million mph. **It must have been a fitting display for the Fourth of July!**

It is the heavy element debris and dust from such explosions along with the existing hydrogen, which collects and condenses into stars, planets, asteroids, comets and anything else roaming around the universe. It is also the origin of all the different elements found today in the universe as well as in our own solar system. In a sense, the universe is one great recycling operation. Supernova explosions produced all the elements (except hydrogen and helium) found in the universe. That goes for all the stuff we are made of too. For example, our blood is red because of the iron in it; this iron was produced in a supernova explosion of a long gone star . . . . as was the gold in the ring you may be wearing.

Approximately 2% of the sun's composition is elements from a previous supernova. In fact, astronomers think that our sun is a third generation star.

That is perhaps worth repeating: all life is made of the stuff of stars that exploded long, long ago. In that sense we are indeed children of the universe.

BLACK HOLES

The final stages of a supernova explosion, described above, may result in the formation of a '**black hole**.' Black holes are the celebrities of the universe because of their peculiar and frightening characteristics. A black hole contains so much mass within such a small volume that the gravitational field produced within it is so great that nothing, not even the photons of light, can escape its pull. Since they cannot therefore be seen, they are called black holes. Not only that, but a black hole often 'feeds' off the gas, stars and even other black holes in its neighborhood, sucking in all that comes near, allowing nothing to escape. Current observations lead astronomers to believe that a black hole exists at the center of every galaxy. It is also thought that a black hole is alive and well at the center of our corporate headquarters.

THE ORIGIN OF EARTH, THE MOON AND OTHER PLANETS

Strangely, we seem to know more about the history of stars and galaxies than we do of our own planet earth. There is no clear agreement among scientists today as to just how the earth, moon and other planets were formed.

The earth is approximately 4.7 billion years old, which is not much different from the age of the sun. It seems certain that the sun and planets all coalesced from gas and dust of previous supernova explosions somewhat simultaneously. The lighter hydrogen gas became the center and started the thermonuclear fusion in the sun, while the heavier elements were left over and condensed into the planets. The early solar system was a dangerous place, it was a period of great collisions and bombardments from comets and other roaming bodies, all orbiting the new sun. Each collision would produce enormous energy releases from the still molten inner planets.

The currently favored view (or 'default theory') is that the moon was formed as a result of a collision (at 25,000 mph) between the earth and another planet possibly about the same size as Mars, part of which fused with the still molten earth and another part flew off and became the moon. Before this impact, the earth had an 8-hour day, after impact its rotation had increased to a 5-hour day. Because of the frictional effect of the tides, the earth's rotation has slowed to the current 24-hour days. The moon is receding from earth at the rate of 1.5 inches per year.

Other planets were formed in much the same way. In addition, there was still a great deal of matter left over, which eventually formed the asteroid belt.

It is often asked: "why are all the stars and planets spherical yet asteroids are not?" The answer is simple: as a body accumulates more and more mass during this 'condensing' period, the gravity it produces eventually becomes so great that it crushes itself into a sphere. Asteroids have not reached that size and remain irregular shaped.

The action is not yet over and there is still a lot going on in the solar system. Asteroids are constantly colliding and when that happens they get nudged out of their neat orbits and may stray to within the influence of earth's gravity, resulting in bad things happening. It was just such an occurrence that caused the mass extinction that saw the demise of the dinosaurs some 65 million years ago. Then, an asteroid the size of Mount Everest or Manhattan hit the earth in the Gulf of Mexico and wiped out all the dinosaurs and about 90% of all animal life. Only with the dinosaurs safely out of the way could mammals emerge to evolve into man and many other animals.

Once the earth's crust solidified and cooled somewhat, life was able to emerge, first as algae, then bacteria . . . . but that's another story . . . . see Part 3, Evolution.

The earth receives warming radiation from the sun. However, despite the current squabbles over 'global warming,' the earth is heading for another ice age in about 15,000 years. Cycles such as this have been occurring with regularity over the life of the planet and our atmospheric mess is not likely to change this; we will have to adapt to a long and very chilly period.

Worse, the earth's core is slowly cooling and in about 2 billion years the fire will go out and it will no longer be able to sustain life. Another reason to be looking around for 'greener pastures.'

PLANETS OF OTHER STARS

The quest to find planets around other stars has long been a goal of astronomers, but progress has always been hampered by the limitations of the telescopes. Once the Hubble space telescope provided the needed power, then the problem became one of relative brightness, for any such planet's reflected light was totally swamped by the sheer brightness of its parent star. This effect is why we can't see the stars during the day.

However, certain stars were found to vary slightly in intensity, in a way likened to blinking. It was correctly suggested that a planet passing in front of the star and causing a brief reduction in light intensity might cause this.

Finally, planets were found not by light but by gravity. Consider a children's playground seesaw. Children of similar weights sit at opposite ends and balance each other. Now, as any parent knows, if an adult wishes to play with the child, he must sit much closer to the pivot point than the child, so that the effect of the adult's larger weight is less effective at the shorter distance and a balance is produced.

Similarly, as a planet revolves around a star, the mutual gravitational force causes the star to make a corresponding but much less noticeable deviation in its own position, it is actually rotating in its own very small orbit. In reality, both star and planet are rotating around a 'balance point,' which may actually be located within the sphere of the star itself.

Now we cannot see this, so called, exoplanet, but by careful observation the star can be seen to 'wobble.' By measuring the amount and period of 'wobble,' the speed and mass of the unseen planet can be established. It is only in the last few years that planets outside the solar system have been discovered but now more are found almost daily.

It is estimated that there are at least 6 billion stars with planets in the universe. This is probably a gross underestimation in view of the rate at which new ones are being discovered.

DARK MATTER

As our knowledge of the universe grew and more accurate observations were compared to theoretical predictions, a troubling anomaly arose. It was first suspected and then proved, that the stars were rotating around the center of their galaxies at speeds that should have thrown them outward out of their orbits. In other words there seemed to be a much greater gravitational effect than could be accounted for by the known mass of the stars.

The only way observations could be reconciled with the known laws, was if there was an enormous amount of matter in the galaxies that could not be seen or otherwise detected. This was dubbed "**dark matter**" and is today considered a fact, although not really understood.

THE END OF THE UNIVERSE (Possibilities #1, #2 & #3)

The following discussion of the possible end of the universe is highly speculative and is very much at the center of research today. Thus the four scenarios summarized here have been very much simplified in order keep the theories 'understandable.' This simplification is justified in that we are merely intending to show how science is constantly attempting to advance our knowledge of the universe. We certainly have not reached the point where we need to say "enough is enough, we give up . . . . God did it." With that in mind, let's look at what the scientists are suggesting *might* happen to the universe.

When a ball is thrown upwards into the sky, the force of gravity acts on it, slowing it down until the ball stops rising and starts to accelerate back to the ground. Naturally, the harder you throw the ball the higher it goes before turning back. If you could throw the ball hard enough (and if there was no air resistance to slow it down) you could eventually throw the ball so hard that the force of gravity would be overcome and the ball would keep going and escape the pull of the earth's gravity.

In a similar manner, the big bang sent the entire universe's matter into space at enormous speed. It was naturally expected that, even though everything was receding from everything else, the gravitational pull of the rest of the universe would act on the various stars and galaxies and, so Friedman argued, one of three things would happen.

1) The **"Big Crunch."** Eventually the expansion would be overcome by gravity and everything would come to a stop and then starting falling back to the center. Somewhat like the ball thrown into the air. This, it was thought, would lead to an enormous implosion essentially equal, but opposite to, the big bang.

2) **"Continuous Slowing Expansion."** Here the universe would slow down and almost come to a stop, retaining just enough outward motion to avoid the 'Big Crunch'. In effect, the universe would expand forever and never quite coast to a stop.

3) The **"Big Freeze."** In this scenario, the universe would continue its steady expansion indefinitely. In this case, eventually all galaxies would lose sight of each other. As the stars and galaxies ran out of hydrogen to burn, the universe would just die away, expanding forever. *A sad end to such a magnificent entity*. However, that was before the Hubble Telescope.

DARK ENERGY

The Hubble Space Telescope has allowed the accurate study and measurements of far galaxies to an extent never before possible and a surprising discovery was made in 1998. The universe, as it expands, is actually accelerating. Not only are galaxies receding from each other but they are receding at *an ever increasing speed*.

This acceleration cannot be explained by the rotational energy of the galaxies or by any other known principle of physics. Astronomers have now formulated a theory that fits this acceleration. This theory suggests that there is an enormous amount of unseen and otherwise undetected energy out there strong enough to overcome the gravitational and dark matter forces. This outwardly acting energy is providing the forces required to produce this acceleration. This force has been termed **"Dark Energy."**

THE END OF THE UNIVERSE (Possibility #4)

4) The **"Big Rip."** The implication of the effect of the newly discovered dark energy is that if the universe keeps expanding forever at an

61

accelerating rate, everything: all stars, galaxies, people, and even atoms would eventually be torn apart into oblivion in what has been called the "**Big Rip**" . . . . in about 20 billion years.

As best the scientists can currently estimate, in order for the universe to conform to the known laws of physics and to behave in the observed manner; then the universe must consist of:

4% visible matter, 23% dark matter and 73% dark energy.

Since (in 2009) we don't really have the slightest idea of what dark matter or dark energy really is, we still have much to learn about the universe.

RELATIVITY

The current scientific knowledge and theory of the universe is nowadays steeped in advanced mathematics, which I, even having a couple of mathematics degrees, find almost impossible to follow, yet alone understand.

It is accepted, however, that Einstein's General Theory of Relativity will continue to gain in validity in the future, so a brief comment is appropriate here:

Space and time are believed (perhaps by most astrophysicists) to be intertwined such as to form one entity, referred to as space-time. Likewise, mass and energy are directly related. These, together with gravity, form the basic parameters of the universe. If this sounds a little vague, that's because it is difficult to put complex mathematically based ideas into simple words.

Is this really surprising though? Throughout history, what once seemed like deep mysteries have resolved into everyday knowledge as scientific discovery has advanced. (Thereby eliminating the need for a god in that area).

Currently the space-time continuum is often likened to a trampoline surface with heavy masses like the earth and sun taking the form of heavy balls rolling around on the surface. These masses put a moving "dent" in the surface and if a smaller mass gets too close, it will roll into the depression and close in on the larger one. This analogy is often used to demonstrate the effect of gravity (the depression) and the warping of time as the rectangular gridlines get bent along the path of the rolling balls. **Don't worry, I don't think it is a very good analogy either, but you get the idea**.

WORM HOLES

Suppose we take hold of one edge of this space-time 'blanket' and fold it over. Let us also suppose that there is a ball on top. If the depression is deep enough, maybe it is possible that a connection between the top surface and the bottom could be opened up. If such a connection were indeed possible, it would allow travel through enormous distances via the short cut or, as it has been called, a "**worm hole**."

Now this is a terribly vague concept to grasp, and to put it bluntly: "way out in left field." But, it is mathematically possible . . . and bear in mind that most of what we know of the universe today was once just "mathematically possible."

Understanding more about the universe today is largely a matter of mathematical theory. Its progress, unfortunately, depends on which school of thought is in vogue and thus receives the lion's share of the available funding necessary for research to proceed. But we'll make progress somehow, we always have!

QUANTUM PHYSICS

This brief review of the universe would not be complete without at least a quick look at quantum physics. Quantum physics is a relatively recent topic in the mainstream of research and deals with the very, very, small sub-atomic particles and the laws governing their behavior.

Not too long ago, not much more than a hundred years, it was thought that atoms were the smallest particles in the universe. It is now known that there are numerous much smaller particles forming the building blocks of the universe.

The study of this branch of physics is extremely difficult for two important reasons: (1) it was soon found that particles of quantum physics follow a completely different and non-intuitive set of laws than conventional physics and (2) the problem is that we are dealing with particles so small that they cannot be observed and the only way research can progress is by studying the effects of the interactions between these tiny particles.

Most of us have heard of the recent, most complicated and most expensive research device ever constructed . . . the Large Hadron Collider or LHC. This device is a 17 mile underground circular tunnel in Europe which houses tubes in which various particles are magnetically accelerated to near the speed of light and then allowed to smash (collide) into particles travelling in the opposite direction. The results are then recorded by instruments which include various forms of cameras, detector plates etc. Scientists can then, hopefully, determine the various properties of these particles.

The collider has only recently been placed on-line and results of any significance have yet to emerge, but it is expected to shed light on the mechanics of the 'big bang' creation of the universe.

Note: as this section is being edited (July 2012), the news networks are excitedly reporting that scientists working on the LHC have announced that they believe (to five decimal points of certainty!) that they have found the much sought-after Higgs boson, dubbed the "God particle" by the sensation seeking press. But so named for no good reason other than its role in the emergence of the universe at the big bang.

SPONTANEOUS CREATION AND ANNIHILATION

A hugely important point for this writer and for the purpose of this essay is that researchers in quantum mechanics have established that pairs of particles with opposite charges can **pop into existence from nothing**, exist for a very short period of time and then 'recombine,' or **annihilate each other out of existence**.

This incredible concept may not fully or even partially answer the age old question 'how can you get something out of nothing,' but I would say it provides a good 'down payment.' In other words, you do not need a god to create something out of nothing. It may be some time before science can elaborate on this, but I'm willing to wait rather than create a god to fill the gap in our knowledge!

Perhaps an even stranger matter that awaits further research is the phenomena called 'quantum entanglement.' Now while this is way beyond the writer's comprehension, it seems there is evidence that events happening in one place on the quantum level can indeed affect events a great distance away. The stuff of sci-fi indeed, except that it may be real. **We have still much to learn.**

THE IMMENSITY OF IT ALL

It is almost impossible for the human mind to get even the vaguest idea of just how just how big the universe is. What follows are some times and distances that boggle the mind. This section, although containing many otherwise dry facts, should be at least skimmed, in order to achieve a sense of just how minute the human world is within the immensity of it all.

But please try to memorize the items in bold face . . . . They make excellent cocktail party material.

Light travels at 186,000 miles per second, or 670 million mph that is **7½ times round the earth in one second**. A light year is the distance light travels in one year . . . about 5.87 trillion miles.

The nearest star (Proxima Centauri) is 4.3 light years away.

Imagine you take off in a space ship, you circle the earth to build up speed and when you see your take off point flashing by as fast as a machine gun fires, then you are at the speed of light and you set course for the nearest star. Sit back and relax, for it will take you over 4 years to get there. And that is the nearest star.

The moon is about one-sixth the size of the earth. It is about 240,000 miles from earth and moonlight takes about 1.1/3 seconds to reach earth.

The earth is about 8,000 miles in diameter. Its circumference is about 25,000 miles.

The sun is 93 million miles from the earth and is 865,000 miles in diameter. Light from the sun takes 8 minutes and 20 seconds to reach earth.

The center of the sun is at a temperature of 20 million degrees. (Fahrenheit or Celsius, take your pick, this is not a text book).

The sun is burning up its hydrogen fuel at the rate of 400 million tons per <u>second</u>. It will be out of gas and die in about 5 billion years.

There are 200 billion (+/-100 billion) stars in the Milky Way (our own galaxy). **It is named after the Greek Goddess Hera, the queen of heaven, whose breast, it was believed, provided the milk droplets in the sky**.

The diameter of the Milky Way is about 100,000 light years.

There are at least 125 billion galaxies in the visible universe.

One grain of sand held at arm's length obscures the light from over 3,000 galaxies.

It is estimated that there are 10^{21} stars in the universe. In long hand, this is 1,000,000,000,000,000,000,000 stars.

If the earth were compressed to the density of a black hole, it would be smaller than a golf ball, but would weigh the same as it does now.

There are at least 100 billion black holes in the universe. Although many astronomers believe there are many more than this.

There is a black hole at the center of the Andromeda galaxy; it is 4 billion times the mass of our sun. Andromeda is on a direct collision course with the Milky Way. Fortunately, it is still 2.5 million light years away.

Our galaxy is right now colliding with two smaller galaxies: the Sagittarius dwarf galaxy and the Canis Major dwarf galaxy. Luckily, the space between the stars is so great that there are very few, if any, actual collisions and so we don't feel any bumps.

The Very Large Array (VLA) telescope (funded by Paul Allen, one of the founders of Microsoft) will be, when complete, an array of many ground-based telescopes all coordinated to work together as one enormous telescope. It will be able to see back in time 13 billion years . . . to when the universe was a mere toddler at only 700 million years old.

The furthest galaxy so far observed is about 10 billion light years away.

It takes 250 million years for the sun to rotate just once around the center of our galaxy, the Milky Way.

What kind of times are we talking about?

EVENT	ACTUAL TIME	SCALE TIME 1 second = 1 year
Big Bang	13.7 billion yrs. ago	434 yrs. ago
Sun and earth formed	4.7 billion yrs. ago	149 yrs. ago
Bacteria / algae: first life	3.2 billion yrs. ago	100 yrs. ago
Asteroid ended dinosaur era	65 million yrs. Ago	2 yrs. ago
Early man	100,000 yrs. ago	28 hrs. ago
Modern man	10,000 yrs. ago	3 hrs. ago
Jesus	2,000 yrs. ago	33 mins. ago
First airplane	110 yrs. ago	2 mins. ago
My birth (1935)	78 yrs. ago	78 secs. ago
NOW (2013)	0	0
My death (life insurance stats)	8 yrs.	8 secs.
Earth incinerated by the sun	5 billion yrs.	158 yrs.
End of the universe (Big Rip)	20 billion yrs.	634 yrs.

Figure 1

Trying to wrap your mind around 7 or 8 zeros is very difficult, so let's simplify the history of the universe by letting **1 second = 1 year**. What we get is a chart (Figure 1) much easier to comprehend. Moreover, we get a much better idea of the incredibly long time since the universe began. (Numbers have been rounded for simplicity.)

What about distances?

For distances, we'll let **1 inch = 80,000 miles**. (That's about 3 times round the earth). That's a lot of compression but as you'll see, it is necessary if we are to get even the vaguest idea of just how big this thing is! Moreover, even that doesn't do the trick.

Dividing the diameter of the earth (about 8,000 miles) by 80,000 we get 1/10th of an inch, which is about the size of this typewritten lower case bold "**o.**"

Likewise, dividing the diameter of the moon (approximately 2,000 miles) by 80,000 we get 1/40th of an inch, which is about the size of this typewritten bold period "**.**" and about 3 inches from the earth.

Similarly, the sun would be a volley ball, (10 inches in diameter), 30 yards away.

So far, it is fairly easy to get a feel for the distances involved . . . . But now:

On the same scale, if you were at the <u>nearest</u> star (another volley ball), these tiny typewritten characters and their 'parent' volley ball (our sun) would be about 5,000 miles away.

But you get the picture . . . don't you? Makes you feel pretty important doesn't it?

ARE WE ALONE?

Of course not!

Suppose there are, were, or will be, others out there, what's the big deal? "They" are not going to look like us anyway, they will be another form of life, maybe wildly different from the millions of life forms we have here on earth.

Regardless, when we do find that there are others out there and maybe see what they actually look like, I do not think it will be the earth-shaking event that everyone seems to expect. If they are sufficiently far away not to influence our day-to-day lives, I think it will be a seven day wonder and will soon fade into the background noise of everyday life. But more on this later.

But what are the chances of finding them? First, we should define what level of life we are looking for. We'll assume that we are talking about intelligent life, capable of communicating, at least in a basic way, face to face, if they have faces. We'll further assume an intelligence level equal to or greater than chimpanzee level.

Granted, there are many basic requirements must be met before life can even be possible. Conditions such as the 'sweet' distance of a planet from its star (not too hot and not too cold); the absence of, or at least, protection from, killer gamma and X rays; the existence of the basic chemicals needed, etc. etc.

Sounds daunting. However, we know that there are more than 10^{21} or 1,000,000,000,000,000,000,000 stars (that's one sextillion or 1,000 billion billion) out there, plus those stars gone and those yet to be born. That allows for a whole lot of misses.

And, when we look at how easily life may be able to start from the right kind of soup, and how far man's ancestors have evolved in just 100,000 years, then to me:

It would seem virtually certain that there are millions of planets with advanced life in the universe.

Communication
Communication, however, is a whole new ball game. For there to be communication, even one-way, the communicants must exist in approximately the same cosmological time zone. A mismatch of 100 million years or so might be acceptable if the receiving beings are 100 million light years away. Of course, in that case, two-way communication would be somewhat slow and tedious. In fact, it is most likely that the originating beings would be millions of years extinct by the time any message or ship containing 'spacedroids' arrives at earth.

Man got his real start after the mass extinction that wiped out the dinosaurs 65 million years ago. This event allowed the evolution of man from the early rodent like mammals to begin in earnest. We've been around, in an intelligent form, known as modern man for just around 10,000 years. But we have only been capable of receiving and interpreting signals from space for around 100 years. That's a bit generous but it'll do.

To expect, or hope, that another communicating life form could exist in the same tiny time slot stretches hope way beyond what I could accept. The window of opportunity is microscopically small. But, you might say, we plan to 'be around' for a while so maybe sometime in the future . . . ?

Yes, but there are serious doubts as to whether man will in fact 'be around' for very long; at least on the cosmic time scale. See comments later.

But OK, *suppose* we send out radio signals at the speed of light; actually we have been sending such signals in the form of stray radio and television broadcasts for the last 70 years or so. By now, those signals will be just 70 light years away and will only have reached a few nearby stars; certainly not nearly enough to have any statistical hope of encountering life.

But OK, just *suppose* there is a suitable exoplanet (a planet orbiting a different star) within this 70 light year distance. If intelligent, advanced beings are there, capable of radio transmissions, we would already know it by listening to our radio telescopes. The chances that life would have developed there in order to reach our level of technical ability just at this time, is to all intents and purposes, zero.

Visitation

Considering the problems with just communication with an alien race, the problems associated with actually swapping visits make such a prospect virtually impossible.

Of course, some kind of future propulsion method capable of near light speed (without the increased mass problem) combined with the use of wormholes or other, as yet undiscovered propulsion, or space-time tinkering methods might change things.

But OK, just suppose . . . .

Overlooking the enormous hurdles of time, distance and technology, let us assume that we (or they) do indeed figure out how to travel there and visit them (or vice versa) through a wormhole, perhaps.

<u>If we are the first</u> to figure out this space travel challenge, and actually make such a journey. We would almost certainly be much more advanced than they, and we would presumably extend to them all the courtesies that the British gave to their empire citizens and the Americans served up to the Indians. After all, we are still human, complete with all the traits we have evolved or developed and, so far, failed to correct. Love for his fellow man has never been man's long suit, and I don't suppose extra-terrestrials would fare any better, after the initial "we come in peace . . ." fanfare.

<u>On the other hand</u>, if a UFO showed up here for real, the boot would be on the other foot and we would have every reason to be scared stiff. They would certainly be more advanced technologically. And morally too, hopefully! But we would have all the bargaining power that the buffalo had (Hollywood and H.G. Wells notwithstanding). In such a scenario, the best we could hope for would be a brief, get acquainted, educational visit. Or, better still, keep out of touch and just observe us with their remote nano-robonauts in what we euphemistically call UFOs . . . . Maybe they are doing just that!

In the meantime, we would be well advised to quit sending them instructions on how to get here.

UFOs

Although not relevant to the primary purpose of this trilogy, this review would not be complete without a mention of the phenomena that has engaged the interest of mankind for centuries: . . . UFOs.

Evidence of ancient or modern aliens
In accumulating notes for this section, it immediately became apparent that the host of reports, investigations and TV documentaries presenting

evidence of earlier visits of co-called ancient aliens are so overwhelming that attempts to do more than list a few of the more puzzling ones here would be useless:

- The ancient precision cut and immaculately fitting stonework in remote places, most apparently beyond the capabilities of modern day technology.
- The massive statuary, beyond the known abilities of the natives.
- The construction and purpose of the Nazca lines, pictorial markings on the ground which only become visible when seen from high in the air. These are known to have preceded the invention of the airplane.
- The worldwide crop circles of unimaginable beauty and purpose.

Most are, by any scientific standard, inexplicable with our present knowledge . . . . but not, however, divine.

UFO sightings

The same is true for UFO sightings. There are thousands of cases of documented and researched sightings by individuals and groups with impeccable credentials. Many astronauts, pilots, scientists and others trained in calm accurate reporting have documented such events while on duty and while being recorded. There are numerous cases where hundreds of citizens have simultaneously observed events, even though the individuals were separated by many miles and unknown to each other. Maybe these phenomena are also best summed up by a bunch of "maybe"s.

- Maybe aliens are indeed here.
- Maybe they have been watching us for hundreds or thousands of years.
- Maybe they have found a method of space travel as speculated above.

- Maybe they are a combination of live and/or android beings.
- Maybe they have unlimited life expectancies, for a possible long journey.
- Maybe it seems that they have no wish to help us.
- Maybe they wish us no harm . . . . so far.
- Maybe they are quite happy, just living "in parallel" with us.

There are many possibilities, ... but divine is not one of them.

CONCLUSION

As a result of this brief review of the universe, I see no evidence of any supernatural influence on the events of the last 13.7 billion years.

'GAPS' NOT FULLY UNDERSTOOD

The cause of the big bang.

The world of the quanta.

The seemingly overwhelming evidence for the existence of UFOs.

PART 3

EVOLUTION

EVOLUTION

BACKGROUND

Back about 40 years ago, one of my hobbies was breeding tropical fish. I concentrated mostly on the lowly live-bearing guppy. I liked the splendid colors in the tails of the males. I also found that the colors, size and patterns could be improved from generation to generation by selective breeding.

As soon as the females could be distinguished from the males, they were separated and raised in a separate tank. When old enough to breed, the best and most handsome male was selected and placed in the female tank. When the resulting young were born, the same procedure was repeated using the male exhibiting the most desired traits. After a few generations, a strain of guppies emerged with much improved size and coloring of tails.

Ironically, as their tails grew larger and heavier, the males lost the speed and agility of their short-tailed predecessors. In the natural wild environment, they would probably have lost out to the more agile short tailed version . . . . survival of the fittest.

In England, my sister Valerie, a retired professional musician, and a lover of collie dogs, took to breeding them for pleasure and show. Within a few generations, she had developed a line that was doing well on the show circuit, achieving champion status for many. Not only that, but her dogs began to have a characteristic look about them that was, to the knowledgeable, recognizable as her 'kennel' look.

What we were both doing was practicing selective breeding, or what might be better called 'un-natural selection' or 'guided evolution.'

FROM ALGAE TO MAN

Almost as soon as the earth cooled enough to solidify its crust, the basic requirements for rudimentary life developed and algae made an appearance.

However, for the next 4 billion years, (fully 90% of the time that earth has existed), nothing much in the way of evolution happened. The earth was settling down from its violent beginnings, oxygen release was starting and other required conditions for life were appearing. Half way, at 2.5 bya, (billion years ago) through this so-called Precambrian era, bacteria and primitive multi-cellular life appeared.

This section briefly examines the history of life on earth from its beginning to the present, 2009. No attempt is made here to catalog or describe the exact path that led to the emergence of Homo sapiens; the whole point is that the entire process proceeded naturally by means of millions of small mutations. In each case the ones best fitted to the challenges of life would survive and generally, but not always, at the expense of less fitted.

The following example is key to understanding the basic mechanism of evolution. **The giraffe did not grow a long neck in order to reach the higher branches of a tree**. It happened that a **random mutation probably produced a giraffe with a little longer neck than others**; thus that animal would have had a slight advantage over the others and was more likely to survive to breed and produce offspring with slightly longer necks, and so on.

Implicit in the word evolution is change, and we tend to think of this as an orderly process. Not so; evolution worldwide is notable for its total lack of order in the manner in which mutations produce changes in the species. It is the success that some mutations enjoy in the face of earthly difficulties that advances the species just a little bit more. This is the driving force behind evolution. **Evolution requires a two step simultaneous event: there must first be the mutation itself <u>and</u> then the need to gain an advantage from it**. A longer neck mutation would be of no advantage where there are no trees, and would be likely to remain a short lived feature.

The time schedule, Figure 3 later in this section gives a condensed sequence of the emergence of the main specie groups. As stated many times above, the underlying purpose is not a technical treatise, but to seek out any evidence of external intelligent influence on the evolution of man. Without any such guidance, man must have evolved solely by a natural process.

Darwin was not able to follow his suspicions to the conclusions that we now accept. His disciplined religious upbringing and schooling must have greatly impeded the necessary mental logical steps to conclude that man evolved without any divine guidance.

This chapter presumes the validity of Darwin's theory on the origin of species, which is now considered an accepted fact, except for pockets of fundamentalist thinking, mostly in the southern USA. Of course, Darwin didn't foresee all the detail now available, but he opened the door to one of the most enlightening branches of science. There is a memorial to him in Westminster Abbey, where Britain buries her monarchs and honors her finest; among the likes of Livingstone, Newton and Watt.

WHAT IS NATURAL SELECTION?

Why do the offspring of any animal grow up to look like their parents? This sounds like a silly question but it has a very definite and important answer. Generally speaking, when male and female animals produce offspring they each pass on information that will determine to a large extent the characteristics of the young. This information is contained in the genes that each parent has, in turn, received in the past from his or her own forbears.

The basic information, such as number of fingers, legs, teeth etc. is available from both parents and is passed on with almost perfect success. Other information such as hair color, intelligence, athletic ability etc. is not so easy to predict and may be taken from either parent, a grandparent or more usually, a mix of previous ancestors. Occasionally, wide differences may occur which seem to bear no resemblance to recent ancestors, these are often (somewhat unscientifically), termed throwbacks. But all in all, the young often follow the traits of the parents.

This genetic information is passed on, we now know, in the form of DNA, an acronym for the molecular structure that contains all the genes coded in such a way as to determine all the characteristics of the offspring. The immediate parents will make the largest contribution to the offspring, but grandparents, great grandparents and so on may each have a diminishing genetic influence on the young.

Left to itself, with no external influence, any species will tend to adapt or evolve the traits which best help ensure its survival. Those that do not will die out.

IT'S ALL IN THE MUTATIONS

Although surely an oversimplification, things so far seem to be eminently reasonable. The interesting part is that things don't always go as we would expect and sometimes the DNA passed on may produce a number of deviations from the 'expected.' Such differences or **mutations** are usually perfectly acceptable and even desirable. Some young may turn out to be much more attractive looking than others, which will probably result, other things being equal (which they never are), in the offspring having a social and procreative advantage over the not so beautiful looking. Thus, the good-looking female will be much more likely to be successful in finding a strong handsome male, should times turn difficult.

Another example would be when one offspring of a wolf, say, is faster and stronger than its siblings; this would mean that it would have a much better chance of feeding itself in cold hard winters.

These advantages are very simple examples of the survival of the fittest, and almost no one has a problem with this type of evolution. The trouble comes when we are told that we all evolved from bacteria or other such unlikely stuff. 'No way,' we hear, 'too many things are much too complicated to have evolved by little accidents.' The most often cited objection is about the eye; 'much, much too complicated' it is said. However what is usually overlooked, are the millions of years of time, and the phenomenal number of generations available for evolution to accomplish these astonishing results. Further it is not just a single line of descendants, father to son, to grandson etc. but parallel lines for as many lines as there are families on earth; all potentially producing mutations simultaneously.

When seen in this frame of reference, evolution becomes much less mysterious.

In addition, a supreme creator would be unlikely to leave large quantities of so-called 'junk genes' lying around, as there are, in our genome. These are genes made obsolete by the march of mutations through the ages. In fact the human genome has great quantities of such obsolete packets of genes no longer of any use to man. These junk genes probably made the research into evolution much more difficult.

Moreover, since mutations tend to be somewhat random, both favorable and unfavorable, evolution goes a long way toward explaining why there are so many defects in the current human and other animal bodies. We can even use the same example used by the evolution skeptics: the eye, which, so I have read, has so many defects and is so grossly inferior in performance to the eyes of many other animals that, if designed by an all-powerful, all-knowing being, would be a clear case of sloppy design bordering on incompetence.

Our eyes, indeed our whole bodies, are no more and no less than what would be expected from repeated small mutations, over many millions of generations, and millions of births within the same generation, all subject to natural selection. Incredible and wonderful, yes . . . . Nevertheless, natural, not designed.

But perhaps Lord Bertrand Russell (1872-1970) (British Nobel Laureate, mathematician, philosopher and peace activist) said it best:

"If I were granted omnipotence, and millions of years to experiment in, I should not think Man much to boast of as the final result of all my efforts."

It is not the intent of this work to educate the reader in the technicalities of evolution, but rather to explain, in reasonable detail, the information that led me to reach the beliefs that I now hold. For those who would like to read more on this subject, a very persuasive and detailed account of how eyes and similar features evolved can be found in "Evolution – The Triumph of an Idea" by Carl Zimmer and "The God Delusion" by

Richard Dawkins; both books treat the subject simply and clearly, and make for easy reading.

NOT-SO-BENEFICIAL MUTATIONS

When mutations are unfavorable and carry forward to the offspring, these individuals will be at a disadvantage compared to the normal individuals and will probably not survive, leaving the original, more fit ones to survive.

There are many examples in the animal kingdom of this type of mutation where, for some reason, the DNA contains anomalies or errors and the young may be born defective in some way. Some dog breeds for example are prone to bone and muscular problems, others to serious eye problems. In the natural wild environment, these offspring do not survive and so the defect is eliminated before it can harm the general population of the species.

However, DNA is not anywhere near as simple as this. Many unfavorable mutations may not exhibit themselves for many generations. Others may survive and invade the species to quite an extent but not sufficient to cause an extinction.

The human race is not immune to these kinds of DNA problems. Fortunately for many, our advanced medical and scientific knowledge allows us to provide these folks with assistance that allows most to live normal lives. There are millions of people alive today that just a few hundred years ago, being born with defects, would not have survived to adulthood. My younger son and I are among them.

It is estimated that the cave men, or more accurately, people of the Stone Age, rarely lived beyond thirty-five years of age, and until a few hundred years ago, the life expectancy of humans improved only because of natural selection. Now the gains are almost exclusively due to healthy food, better life styles and improvements in medical practice. It seems to me that natural selection has almost been arrested.

The enormous advantage that the human race enjoys is the evolution of its large brain with much greater reasoning powers. This has enabled it to overcome many of the missteps of unfettered natural DNA problems.

HOW NEW SPECIES EMERGE

Occasionally a species may become physically or geographically separated into two or more isolated groups. In such cases and because of the random nature of mutations, the groups will slowly but inevitably evolve along different lines until they are sufficiently different so as to be classed as new or distinct species. Bear in mind, however, that 'species' is a man-made classification and there is no biological absolute point at which differences warrant being classed as a new species. Indeed, biologists are constantly tinkering with the biological requirements to be named a species.

It was these kinds of puzzling facts that so fascinated Darwin in the Galapagos Islands. There are several well-separated islands where birds, which were clearly one species originally, were separated, perhaps by a hurricane, onto different islands. The groups evolved independently over the ages and are now well-defined and clearly separate species. Darwin wondered how this could be

. . . . And now we know the rest of the story!

The same kind of thing has happened all over the globe. Modern man for example, being a mobile species, spread (probably from Africa), to all parts of the world. There being no jet planes, trains, cars or even bicycles then, there would have been virtually no interbreeding among the widely separated groups and so the different groups evolved into the many different races that we see in the world today.

RECOMBINATIONS

Nowadays, however, the pendulum is swinging back. The ease and speed of modern travel and relocation is facilitating easier inter-racial breeding. Thus differences in appearance (and even personality traits) are becoming blurred.

In the future, now that we have relatively cheap, easy and plentiful travel and relocation opportunities; interbreeding among groups of different locales is likely to continue to grow. This is especially likely as mixed race unions continue to grow in social acceptability.

Soon we'll all look the same which might go a long way toward eliminating many social problems.

Groups that become separated from the main line of any species will experience a whole new set of mutations which may consolidate into a new species. If the two groups (now different species) should meet up again and intermix, it is possible that one of the groups will become re-absorbed into the other. There is a school of thought that believes this is what happened to the Neanderthals who are known to have adapted to the cold quite well, were quite intelligent and, it is believed, had a rudimentary language for communication.

DEAD ENDS AND DIE-OUTS

Species that emerge and later prove to be unable to adapt to changing conditions, will become extinct.

NATURAL EXTINCTIONS

Many times in the history of the earth, an event has occurred that has had disastrous effects on numerous species. Asteroids, super-volcanoes, tsunamis, ice ages and diseases have all, at one time or another, resulted in the extinction of many species. Geologists agree that there have been five major mass extinctions throughout the evolution of life on earth. Any one of which would likely have eliminated the human race had it been around at that time.

It is estimated more that 95% of all species that ever lived on earth are now extinct.

The most well-known mass extinction was probably caused (along with other contributing factors) by the large (mountain sized) asteroid that collided with the earth some 65 million years ago. It wiped out all land animals living above ground including the dinosaurs; and in so doing, opened the door for man to evolve from the small shrew-like, underground mammal that survived the asteroid's aftermath.

UNNATURAL EXTINCTIONS

Man has probably been the cause of many of the unnatural extinctions on earth. There are many examples where man has hunted, killed and driven a species into extinction. The dodo bird is a classic example of such a tragedy. A search on Google brought up the following summary by David Reilly:

> *In the year 1598 AD, Portuguese sailors landing on the shores*
> *of the island of Mauritius discovered a previously unknown*
> *species of bird, the dodo. Having been isolated by its island*
> *location from contact with humanity, the dodo greeted the*

new visitors with a child-like innocence. The sailors mistook the gentle spirit of the dodo, and its lack of fear of the new predators, as stupidity. They dubbed the bird "dodo" (meaning something similar to a simpleton in the Portuguese tongue). Many dodo (sic) were killed by the human visitors, and those that survived man had to face the introduced animals. Dogs and pigs soon became feral when introduced to the Mauritian eco-system. By the year 1681, the last dodo had died, and the world was left worse with its passing.

Attempts to arrest this kind of unthinking destruction of life have recently been helped somewhat by acts such as the Endangered and Threatened Species Act in the USA and other modern nations. However, much of civilization has yet to recognize the value of life in all its forms. Man seems to be just about the only animal that destroys life (of any kind) for pleasure or fun. A genetic defect we would do well to seek out and eliminate.

INSTANT (ALMOST) EVOLUTION

But what about life forms of very short life cycles? There are two whole aspects of evolution that are available for everyone to observe and from which to learn. One is the very small-scale creatures with very brief life cycles, such as the fruit fly. The second is the world of bacteria and viruses.

Fruit flies are (or were, when I was younger) the mainstay of the biology labs devoted to the study of evolution. They have extremely short life cycles and thus the impact of mutations over many generations can be observed in a relatively short period of time. Where a single human generation would take around 20-25 years, the fruit fly manages it in a day or two. Thus, the fruit fly is ideal for experimentation, whereas genetic tinkering on humans is time consuming and generally frowned upon.

Bacteria and viruses have even shorter generation cycles and, although not particularly animal-like, they exhibit most of the classical signs of mutation and evolution. Anyone who goes for a 'flu shot today and takes the trouble to ask, will be told that it is not the same shot that they got last year. Should they delve even further and ask why, they will be told that it is different because the 'flu virus of last year has benefited from natural selection and in evolving, has mutated into a strain of the virus that is resistant to the killer drug administered last year.

Luckily for us, the pharmaceutical companies have been hard at work all year developing a new killer drug with which to inject us. Thereby we have maybe one more year of breathing space. And so the battle rages. Barely a day goes by without stories in the news of outbreaks of the latest mutation of the flu virus. **One day we may not win**. There have been many pandemics over the years that have killed sizeable percentages of the population.

It always amazes me that in the USA, the most technologically advanced nation in the world, and in the face of elementary examples like the above, there are apparently millions of folks who deny the truth or even existence of evolution.

WHAT OF THE MISSING LINK?

The so-called missing link seems to me to be a myth. It is said that we have never found a direct link between man and say, chimpanzees. Of course not, they are the end results of two completely separate lines of species evolution. They may share 98% common DNA and some individuals may look and act suspiciously alike, but that is because we had a common ancestor, millions of years ago.

Trace modern man back through the eons and trace the chimpanzees back through the same time and you will arrive at a point where their common ancestor split into two distinct species, which probably looked somewhat reminiscently similar to modern man and chimpanzees, but both species then embarked on the long journey of evolution into very different animals. It is after the split that man evolved all the attributes that separate us from other species. To wit: the development and use of language, a much larger brain, followed by the ability to use sophisticated tools and make clothes which, in turn, maybe reduced the need for body hair, etc. etc . . . The chimpanzee's ancestors may have taken to the trees for whatever reason, developing longer and stronger arms and use of the tail, and so on. Or maybe it was the other way round, and we lost the use of a tail. Either way it matters little for the purpose of this essay.

No, the only link would be the common ancestor from which two very different species evolved. Man did not evolve from chimpanzees, although modern man's behavior might sometimes suggest otherwise.

MAN'S CLIMB UP THE EVOLUTIONARY LADDER

Simplified Chart of Typical Evolution Patterns

Below is a <u>very</u> simplified schematic chart showing the manner in which evolution progresses.

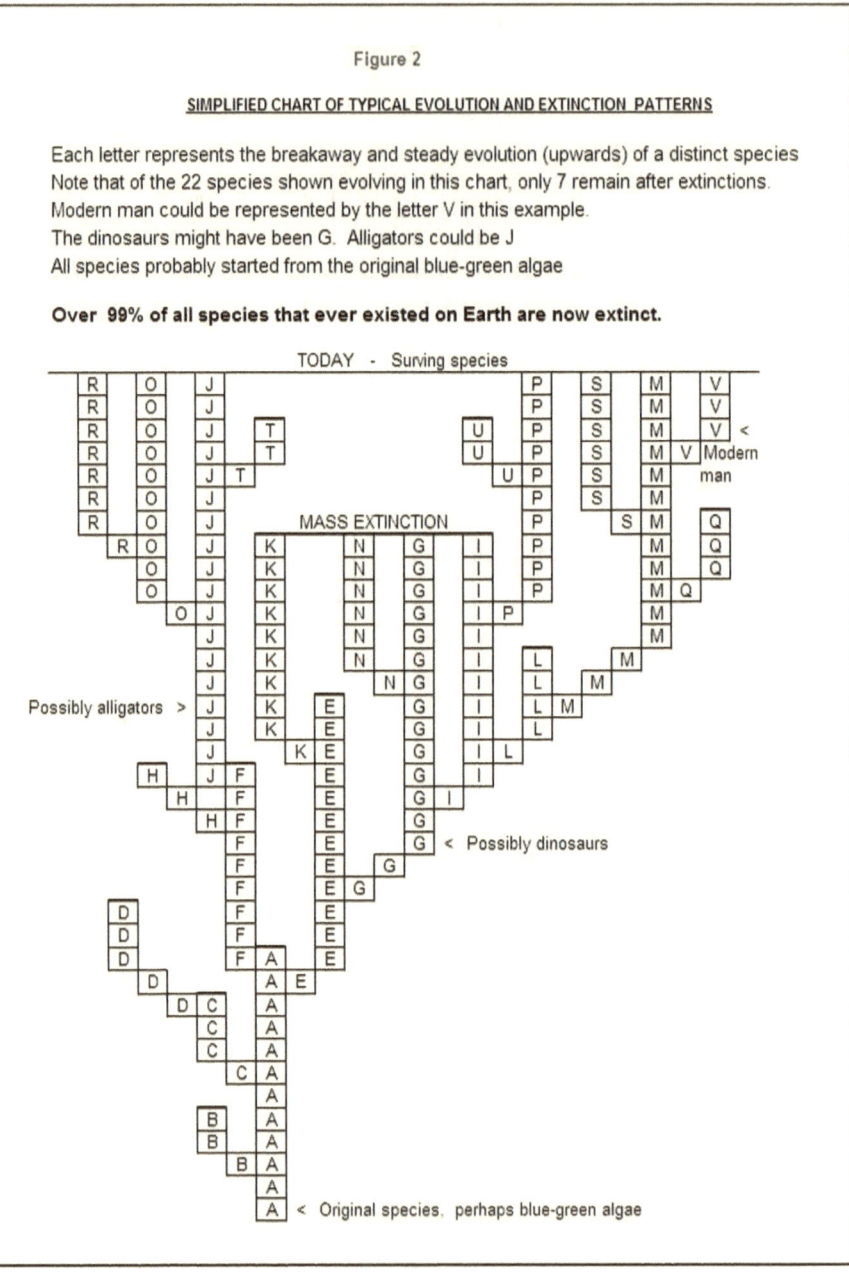

Figure 2

SIMPLIFIED CHART OF TYPICAL EVOLUTION AND EXTINCTION PATTERNS

Each letter represents the breakaway and steady evolution (upwards) of a distinct species
Note that of the 22 species shown evolving in this chart, only 7 remain after extinctions.
Modern man could be represented by the letter V in this example.
The dinosaurs might have been G. Alligators could be J
All species probably started from the original blue-green algae

Over 99% of all species that ever existed on Earth are now extinct.

The notes on the chart are pretty much self-explanatory, but a few comments may help.

Read the chart from the bottom upwards in time. The first living species to emerge is 'A' which continues to evolve with time (upwards) in the chart. Thus, the higher 'A's are more evolved than the lower 'A's.

After some time, mutations have occurred such that a new species 'B' is formed which breaks away from the 'A' line. Unfortunately for 'B,' and for whatever reason, it becomes extinct in a relatively short time.

Later, a further breakaway species 'C' occurs, which in turn mutates into a different species 'D.'

Occasionally different species will interbreed and may produce a third species. I would expect that in such cases the DNA of the two lines would have to have more than a few common characteristics, for the offspring to be fertile.

But back to the chart: Note that higher in the chart, species 'K,' 'N,' 'G' and 'I' are all eliminated in a mass extinction.

The chart is just a simplified example of how evolution occurs through time. In real life, however, the 'tree of life' is much more complex. For example, there may have been far more than just one initial species and new forms of life may be springing up even today. The whole story of life on earth is a very fluid, ever-changing scenario. It happens so slowly, with changes occurring only once per generation, that we don't see the progress in species that have long gestation and growth cycles.

. . . .

In the chapter on the universe, I described how the earth was formed. We take up the story from when the earth's crust solidified some 4.7bya (billion years ago):

A Time Chart of the Evolution of Life on Earth

Below, I have included a similarly very simplified time chart covering life on earth from the earth's formation 4.7 billion years ago to the present day. The intent here is to impress on the reader the incredibly short time that man has been around in this universe.

Figure 3		
FROM THE BIG BANG TO MAN - (13.7 bya to the Present)		
ACTUAL TIME (Actual years ago)	**SCALE TIME** (1 sec.=150 yrs.)	**EVENT**
13.7 billion	3 years ago	**Big Bang**
4.7 billion	1 year ago	**Earth's crust solidifies.** **Collision with 'rogue' planet creates the moon.**
3.9 billion	10 months	**Blue-green algae - the first sign of life.**
3.6 billion	9 months	Bacteria, single celled organisms.
2.5 billion	6 months	Oxygen release starts - cellular life.
540 million	6 weeks	"Cambrian explosion" of new species evolves.
438 million	4 wks 6 days	Extinction occurs. 50% of animal species.
248 million	19 days	Mammals, crocodyloformes, reptiles, turtles.
230 million	18 days	**First dinosaurs**
208 million	16 days	Age of dinosaurs; first birds and mammals;
146 million	11 days	Dinosaurs dominate the land, crocodiles.
65 million	5 days	**Asteroid hits earth. All dinosaurs extinct.** **Age of mammals begins; primates.**
54 million	4 days	Modern birds and mammals; giant birds, whales.
38 million	3 days	Browsing mammals; sabertoothed tigers.
10 million	1 day 15 hrs.	Apes in southern Europe, ramapithecus
5 million	19 hrs 30 mins.	**First hominids (australopithecines).** **Brain 1/3 size of modern humans**
2 million	3 hrs 55 mins.	Megalodon, modern form of whales
1.8 million	3 hrs 32 mins.	Widespread glacial ice (ice ages)
100,000 years	12 mins.	**First humans (homo sapiens).**
10,000 years	1 min. 10 secs.	**Modern humans**
2013 yrs. ago	13.4 seconds	**Jesus born**
1979 yrs. ago	13.2 seconds	**Jesus died**
In other words: if the earth is one year old then . . . . . **Jesus was born 13.4 seconds ago and lived for just 1/5 of a second.**		

Note that the chart is not to scale. For instance, the top of the chart covers the first 10 billion years, whereas the bottom covers only 10 thousand years.

The 'ACTUAL TIME PERIOD' column and the 'EVENT' column are self-explanatory. However, in order to provide an idea of just how briefly man has been on this planet, I have added the column 'SCALE TIME PERIOD' that gives the relative times **as if the earth were formed one year ago**. This is a time compression of 150 years into 1 second.

What struck me, as I read and thought about the sequence of events of the last 4 billion years, is the utter randomness of it all. The evolution of all life has been inextricably bound up with the geological evolution of the earth itself and upon extraterrestrial events such as asteroids, meteors, etc. It seems that the simplest of unrelated events has made vast differences in the outcomes. In fact, the only thing that makes any order out the turmoil is the law of natural selection. Without that, there would be no progressive evolution and life would be just a jumble of brief, primitive existences, followed by quick extinctions.

Despite natural selection, scientists identify five major mass extinctions and at least seven minor ones. Of course, how one defines major and minor extinctions is important. I could not find a technical definition but I would say that a major extinction results in the loss of a sizeable percentage of the earth's fauna (and flora); while a minor extinction results in many species being lost due to an event that had no relation to the species involved.

Regardless, evolution has finally produced Homo sapiens, a species that has reached a level of intelligence that no other animal has achieved (at least that we know of). This intelligence and reasoning power has led to man being able to think ahead and so provide a measure of control over his future, to an extent not available to other animals.

Despite the heading of this section, man did not climb the tree of life, nor did he use a ladder. He evolved . . . . naturally.

DO GENETICS HAVE ANY EFFECT ON OUR DAILY LIVES?

There is a whole lot going on in the gene department that we do not see. There are millions and millions of little mutations going on all the time; they are so small that we do not notice them.

An obvious example is the speed and apparent ease with which the flu viruses mutate and adapt to defeat the drugs we take to fight them. New drugs have to replace last year's to maintain our resistance to the new hostile strains. Maybe that is getting a little too technical for the purpose of this work; sufficient to say that evolution is a constant force to be reckoned with.

THERE IS MUCH EVIDENCE OF UN-NATURAL EVOLUTION

Nowadays, because of the slowness of natural evolution, improvements in the human condition come more from medical advancements than from evolution. Better nutrition and training accounts for better athletic performance, and new world records are almost never a result of natural evolution.

For example, tallness is an important characteristic for basketball players, but basketball players are not in competition with other folks for survival, and so tallness remains a relatively random trait. If survival depended on tallness, then the human race as a whole would gain height. This, of course, is exactly what happened to the giraffe with its evolved long neck and ability to reach the higher branches for food.

Even gains in our life expectancy seem to be the result of better medical knowledge and infant survival. Natural evolution of man continues but at the (relatively) extremely slow pace that it has always moved. I cannot think of any evolutionary change directly attributable to natural selection in the last few hundred years.

Are our babies born any bigger? They would be, only if large baby size was an advantage in survival. Since large baby size does not seem to be related to survival rates, I wouldn't expect babies today to be much different than babies of hundreds of years ago.

There is no reason to assume that evolution is finished with us, although we seem to be giving it a hard time in its attempts to "improve" us. For example, the medical profession is, in a manner of speaking, trying to halt the less desirable side of natural selection. We do our utmost to keep alive any human being, no matter what defects are present at birth. We are most proud when we overcome these defects and provide whatever assistance technology can muster to ensure that the person survives to live a normal happy life and, of course, to bear young. Even though the offspring may, in turn, carry the same problem genes. This, no matter what its moral or social benefits, is not natural selection.

However, millions of us, myself and my younger son included, would not have survived to bear young had it not been for the advancements in medical science. Since we are now both reasonably productive and peaceful members of society, I guess we score one for the medics and for the benefits of civilization. Nonetheless, the gene quirks that nearly cost us our lives are probably alive and well in our genome, hidden for now, but perhaps ready to resurface in our future descendants.

It is well known that genetic defects that are allowed to survive and gain a footing in the human genome are unlikely to disappear of their own accord. They will reappear from time to time, causing much grief. Such is the price of a high infant survival rate. **As far as humans are concerned, civilization and its medical technology is clearly the enemy of natural selection.**

SUMMARY

From this brief, layman's study, it seems clear to me that man has evolved in the same manner that all other life on earth has done, that is, by natural selection within the environment in which he found himself and subject to the effects of the earth's many upheavals.

I see no scientific evidence of any 'intelligent design' or supernatural influence on the events of the last 4.7 billion years.

'GAP' NOT FULLY UNDERSTOOD

The initial transition from chemical to reproductive life.

PART 4

WHAT ABOUT THE FUTURE?

WHAT ABOUT THE FUTURE?

CAN MAN EVEN SURVIVE FOR LONG?

Before we take a look at what man's future might be, it may be worth a quick check to see whether we will even be around very long. By 'very long' I mean a great deal longer than the puny 10,000 years modern man has already clocked up on earth. A lot can happen in a billion years or so.

Consider two chilling facts:

There is nearly universal agreement among scientists that more than 95% of all species that ever lived on earth are now extinct.

Man is the only animal that has ever developed the capability to destroy its own species.

I suggest that these two facts alone do not auger well for man's future. There will certainly be more natural mass extinctions, like the several that have occurred already. Further, we are not well equipped or prepared to face nature's wrath for:

In terms of cosmological size we are virtually non-existent.

We live in a world where everything is 'manageable' so to speak. By that I mean we rarely have reason to look beyond our local environment. Mountains are about as big as things get and unless we are scientists looking through a microscope, then ants are about as small as things get. We become comfortable in this environment.

However the largest building on earth can barely be seen from a window of a jetliner at 50,000 feet, and the entire earth could not be seen from the nearest star.

We are frail in the extreme

Without special protection we cannot survive outside of a tiny temperature range of about 100 degrees; in a universe that ranges through millions of degrees. And, we can breathe just one specific mixture of gases.

We are not strong enough to stand upright on any solid planet the size of Jupiter, indeed we would be crushed by the gravity. We could not survive if earth were only half the distance to the sun or three times further away; the so-called Goldilocks distance, not too hot and not too cold.

We are the 'new kid on the block.'

As far as time is concerned, modern man has been around for about 10,000 years. Not much to write home about in the 3.7 billion years of life on earth, or the 13.7 billion years of the universe. And we have yet to be tested by a real cosmic event.

In other words we have evolved to be creatures suitable for just one purpose: life on earth. Take away any of these 'environmental conditions' and we become extinct. Immediately.

Now catastrophic events may not eliminate us entirely; just return the few of us that may remain to start over again, perhaps from hunter-gatherer status.

So, just what might break our happy little bubble?

The following possibilities range all the way from certain extinction to those causing a major set-back of the human race They are listed in order from what I believe may present the most imminent practical and realistic threat to man's status quo, (but not necessarily full extinction), all the way down to events that, in all likelihood, will never occur.

(1) Pandemic

The main fear here is not a fast spreading illness that would take its toll and peter out as was the case with the Black Death in Europe in the Middle Ages, or the Spanish Flu epidemic at the time of WWI, serious as they were.

No, what is feared is a new, unknown deadly virus that could sweep through the world at lightning speed, or at least the speed of a modern jet airliner, leaving no time for man to develop an anti-dote. There have been several 'scares' recently of various viruses which have popped up in the developing parts of the world and have a resistance to all existing drugs.

If a jihadist group ever got its hands on such a viral weapon, it would be quite likely to use it, with or without realizing the consequences of its actions. Such is the idiocy of all religious extremes.

Such a virus could free the world of nearly all of its human population in a few days. Certainly it could be a major setback to mankind.

(2) Nuclear war / accident

There is an enormous amount of nuclear material left over from the cold war, which has neither been made safe nor disposed of. Much of it is in nuclear devices just waiting to be armed, aimed and fired.

Many believe that it is only a matter of time before an accident occurs, resulting in retaliation to a perceived attack, and with horrendous loss of life. These modern devices are hundreds of times more devastating than the atom bombs of WWII.

There are many terrorist groups and rogue, unstable nations that would love to have nuclear weapons. They would not likely hesitate to use them on their enemies in their 7th century-style religious vendettas

Ever since the first atom bomb was dropped, there has been much speculation that, as even more destructive weapons were developed; a third world war would bring the end of mankind. Although worldwide devastation would surely occur, man would survive, albeit without the technological benefits we enjoy today.

Two quotes are notable:

"If the Third World War is fought with nuclear weapons, the Fourth will be fought with bows and arrows."

. . . . Lord Louis Mountbatten.

"I do not know how the Third World War will be fought, but I can tell you what they will use in the Fourth . . . . rocks!"

. . . . Albert Einstein.

Maybe 'bows and arrows' or 'rocks' is a bit of an exaggeration and such a nuclear exchange would be unlikely to cause a total extinction but would certainly kill or maim billions and would render parts of the world uninhabitable for many decades. In the aftermath of such an event it is unlikely that space travel would be very high on the world's priority.

(3) Mega-volcanoes

Yellowstone Park is a large caldera formed by ultra-large volcanic eruptions, which have occurred many times in the past, with an average frequency of about 600 thousand years. The last one occurred 630 thousand years ago, so on the face of it, we are 30 thousand years overdue for a catastrophe! However, various scientists have reported that "there is no immediate likelihood of an eruption any time in the foreseeable future." I'm not sure what that means but it is not too reassuring if we plan to be here for another million years.

Krakatoa, a volcanic series of islands in the Sunda Straight, Indonesia, erupted and exploded in 1883, causing massive tsunamis and killing at least 36,417 people, while simultaneously destroying over two-thirds of Krakatoa Island. The explosion is considered to be the loudest sound ever heard in modern history, with reports of it being heard up to 3,000 miles (4,800 km) from its point of origin. The shock waves from the explosion were recorded on barographs around the globe.

The eruption was equivalent to 200 megatons of TNT . . . about 13,000 times the nuclear yield of the bomb that devastated Hiroshima, Japan during World War II

. . . .

The above two events occurred at what we might call weak spots in the earth's surface. Such locations exist where the hot magna of the inner earth comes near to the surface crust and occasionally bursts out, with dire results. The result of an event bigger than these two examples (by no means out of the question) could be a world-wide sized dark cloud obstruction of the sun, acid rain and a noxious atmosphere. Most if not all plant life could perish with most animal life to follow. The severity would depend on the size of the eruption, but possibly world population could fall to 10%. Perhaps an ice age might follow, as the greenhouse balance would be disturbed. Not good . . . . **and we're 30,000 years overdue for it!**

Local and isolated groups would probably survive but many of the benefits of cultural evolution (technological and agricultural development) could be lost and man might find himself back in the Middle Ages . . . and much less prepared to survive the hardships than were the actual citizens of those times.

There is no reason to believe that these examples are anywhere near the worst that the 'under-world' could throw at us. It wouldn't take much more to match the effects that followed the Gulf of Mexico asteroid that wiped out the dinosaurs and all above ground animals 65 million years ago.

(4) Large asteroid (or meteor or comet)

It is well known that asteroids can present a threat to the world. Numerous Hollywood movies have made the population well aware of the possibilities, mostly inaccurate. Nevertheless, the long-term threat is very real.

The solar system condensed out of remnants of a supernova explosion of a burned-out star. The main bulk of material formed the sun, lesser amounts produced the planets and much of what was left condensed into a somewhat sloppy accumulation of various sized, rock-like bodies orbiting around the sun, called the asteroid belt. That's a bit oversimplified, but it's close enough.

Every now and again, a couple of these rocks collide and nudge each other out of their orbits. The new path of such an asteroid may put it on a direct collision course with the earth. This happens all the time, but most are very small and burn up in the atmosphere . . . often seen as 'shooting stars' or meteor showers. Unfortunately, there are rocks of all sizes out there and it is the larger ones with which we need to be concerned.

In the early years of the solar system there were many such bodies floating around, some as big as our present planets. The number of large collisions eventually subsided as they collided, were absorbed or demolished each other. Somewhat like the action in an auto 'demolition derby' which eventually dies out.

It is believed that, long ago, a very large body about the size of Mars collided with and merged with the still molten earth; a large fragment was cast off and became the moon. The mutual gravity holds the earth and moon in orbit around each other.

Asteroid collisions still remain a threat to the earth. A year or two ago, a large body (discovered by, and named for the astronomers Schumacher and Levy) was caught in Jupiter's gravitational field, breaking into several large pieces as it descended and collided with the gas giant. NASA filmed and published the spectacular series of earth-sized collisions on the surface of Jupiter; any one of which would have completely destroyed the earth. It is believed that Jupiter has served us well as a shield in the past, and hopefully will continue to do so.

Even more recently in 2004, astronomers identified an asteroid that had a 4.7% chance of colliding with earth in 2029. Fortunately, later measurements and calculations have eliminated that chance.

There are numerous ideas and methods being developed to destroy or deflect such unwelcome visitors. The very large ones will be seen many years before collision, but it is doubtful whether we would be able to avoid the destruction of all life. Luckily, such large ones are now rare in the solar system . . . . **but they are there**.

It is possible that we may be able to protect ourselves from small and medium sized ones by one of the methods now being evaluated; various ingenious schemes of pushing or pulling them out of their paths seems to offer the most promise.

The asteroid/meteor that left Meteor Crater in Arizona was only about 150 feet in diameter but was traveling at about 30,000 mph. It would have had a major effect on all life for hundreds of miles but did not cause any major extinction. Had it landed in water, the ensuing tsunamis would have been a different ballgame!

The asteroid that hit in the western Gulf of Mexico, some 65 million years ago was the size of Mount Everest, approximately 6 miles in diameter, and resulted in the extinction of the dinosaurs and all above-ground life except small, subterranean shrew-like mammals from which all present day mammals (including us) have evolved. The damage resulted from a combination of side effects: an enormous tsunami immediately, followed by the atmospheric pollution and darkening of the sky for years, leading to the worldwide loss of vegetable life and starvation of animal life.

An asteroid of 10 miles in diameter would render man extinct, except perhaps for those in a subterranean, waterproof, self-sufficient bunker, good for perhaps several years of isolation. Just where they would acquire the food and supplies to restart life above ground is not clear.

(5) Sudden disastrous geophysical changes

There are many possible ways in which man's existence on earth may be cut short, or severely compromised. All of the following are quite possible, although their likelihood is a matter of some conjecture among scientists. Just four are listed here with no comment, but a trip to Wikipedia will provide plenty of scary information:

- Changes in the sun's intensity or magnetic field.
- Solar, electro-magnetic sun storms sweeping over an unprepared earth.
- A change in the earth's elliptical orbit around the sun, perhaps caused by a large asteroid near miss.
- Sudden magnetic polar shifts that have occurred regularly in the earth's history.

(6) Unpredictable but probably long term events

Our watery rocky ball floating lazily around the sun every year is enjoying a time of relative serenity. Yes, we have 'terrible' winters and 'blistering hot dry' summers etc. etc., but by and large things are pretty benign. It was not always thus; nor will it be in the future.

- A shift in the earth's axial tilt. The so-dubbed 'wobble' in the earth's axial rotation.
- The incineration of the earth as the sun swells as it runs out of fuel and dies.
- The effect of the moon slowly receding from the earth.
- The effect of the earth's rotation slowing down, due to the dissipation of energy by tidal friction.

Such events may cause the planet to be uninhabitable for life and man would be in trouble if unable to adapt to the changes. Many of these potentially life-threatening events have causes and effects that are not readily apparent to us humans, who currently experience only the short-term effects of daily and seasonal cycles.

(7) A Gamma Ray Burst

A gamma ray burst usually results from an extremely large supernova explosion or collision of two neutron stars. During the life of a massive star, nuclear fusion converts lighter elements into heavier ones.

When the star runs out of fuel, the heat and pressure-generating fusion reaction no longer generates enough outward pressure to counteract gravity; the star rapidly contracts and then explodes. During this final collapse, energy is released along the axis of rotation and forms a highly directional burst of intense gamma-rays. Although these bursts can last from milliseconds to nearly an hour, a typical burst lasts just a few seconds. The initial burst is usually followed by a longer-lived "afterglow" emitting, at longer wavelengths, ultraviolet, optical, infrared, radio and X-rays.

A typical gamma ray burst releases as much energy in a few seconds as the Sun will in its entire 10 billion year lifetime. They are extremely rare: a few per galaxy per million years. Fortunately all observed GRBs have originated from outside the Milky Way galaxy and have been so far away as to be harmless to us.

However, it has been hypothesized that a GRB within the Milky Way, if emitted in our direction, could cause the extinction of virtually all life on Earth. Should a GRB reach the earth it would immediately destroy the ozone shield, which protects the earth from harmful radiation. All living cells would die within one month, except for bacteria deep underground.

The result would be a return of earth to the days where only bacteria lived. **No gamma ray burst has ever affected the earth.**

(8) Black Holes

The creation of black holes results when a very large star runs out of hydrogen fuel, shrinks and eventually explodes in what is called a supernova explosion (see description under Gamma Ray Burst above). What is left condenses and develops a gravitational field so strong that nothing can escape, not even light; hence the name black hole. Black holes 'feed' on anything that comes near them: stars, planets and even smaller black holes . . . whatever.

It is believed that there is a large black hole at the center of every galaxy, but these are generally considered well chained and harmless. However: the ones to fear are the renegade or rogue black holes wandering around the universe at will, snacking on stars and smaller black holes.

If we were to see one coming (perhaps from the blacking out of stars in the line of sight), it would do us absolutely no good . . . it would happily swallow anything we sent up to greet it!

However, an encounter with a black hole is so far in the future that it is not a significant threat, for the purpose of this work.

. . . .

There are other nasty possibilities lurking in the future of course, but the ones listed above should get the message across. Now not all of the above scenarios would result in the end of the earth, or even the complete extinction of the human race. Any one, however, could result in a major setback, possibly pushing man back to a much earlier stage of cultural (knowledge) evolution.

This is most likely to occur when an event results in a partial extinction but leaves pockets or localized groups of survivors to pick up the pieces and carry on as best they can. There are many entertaining movies depicting such scenarios.

Man has already reached a level of 'knowledge' evolution where no man or group of men can know everything needed to preserve the quality of life that we now enjoy.

I have always felt that our current standard of living, and the knowledge and technical infrastructure needed to maintain it, is dangerously fragile. Even before the industrial revolution, specialization was essential to the improvement in quality of life. The thread of necessary knowledge is so long and complex that even the specialist cannot know even one thread entirely. Thus, the destruction of a class of industrial knowledge or scientists could result in partial or even total loss of that specific capability.

The microchip industry, for example, is currently a heavily concentrated and specialized business, and not inclined to share its knowledge easily. Just about any device these days is dependent on microchips. Without these silicon chips just about all of our modern day devices would become unserviceable. Fortunately the technical knowledge to produce these highly sophisticated devices is becoming more and more geographically and nationally diversified. Thus, the potential for a paralysis of technology resulting from such a loss is diminishing with time.

It is often the unforeseen things or 'the law of unintended consequences' that bite us. Ask Buzz Aldrin about what could have happened after something as simple as a switch lever broke off while sitting on the moon. (He used a pen to poke inside and throw the switch that started the return rocket!) "For want of a nail, the kingdom (could have been) lost."

. . . .

There is one other area of major concern to many future-thinking scientists:

ARTIFICIAL INTELLIGENCE / COMPUTERS

The well-known unscientific, but so far still pretty accurate 'Moore's Law' states that computing power roughly doubles every 18 months. Computers have already passed the human being in simple but fairly deep reasoning power. As for example the defeat of the then world chess champion, Gary Kasparov, in a match a few years ago; at the 'hands' of a computer built and programmed by IBM. Ironically, IBM felt that this accomplishment was not that big a deal and shortly after, dismantled the device and the team that developed and ran it.

If computers' performance doubles every 18 months, what will they be capable of in, say, just 100 years? According to Mr. Moore their computing power will have increased by 2^{67} times!!

(For the mathematically inclined: there are 67 periods of 18 months each in 100 years, thus our computers will have doubled in power each time, going from 100% to 200%, to 400%, to 800%, to 1600% etc. 67 times!!).

This is hardly a realistic number, but considering that they are already our equals in power, if not in our reasoning ability, then we have every reason to be nervous. Especially when we realize that it is a one horse race . . . for man is not getting any more intelligent.

Also, much research progress is being made in the area of 'fuzzy logic.' This might be called 'intuition' or the ability of computers to rationalize beyond the strict limits of pure logic. Recent sci-fi TV programs have used the term with some scoffing but this branch of mathematics, (unknown in my day at college) has enormous potential.

The good news is that computers and artificial intelligence have and will continue to generate undreamed of benefits for mankind in just about every conceivable walk of life.

The downside is not so clear. The fear is not that computers will far exceed the human brain in computing power, (they already have), but that they will be able to use this fuzzy logic to pass the point of needing human direction and support. I believe it is almost certain that they will eventually be able to self-replicate and increase in intelligence, and quite possibly be capable of seeking or manufacturing their own energy source. Not far behind will be their self-preservation 'instinct.' There have been several excellent sci-fi movies depicting this scenario.

Now, all these screenplays were written by humans . . . for prime-time human audiences. So of course, they all had happy endings. I'm not convinced we would fare so well in a real situation. Also, bear in mind that none of these movies venture beyond a few decades into the future.

As one scientist wryly remarked . . . "We'd better know where the 'off' switch is."

Well, OK, but that's way off in the future, right? Not so, remember that:

The airplane was unheard of just 120 years ago.

We went from the horse and cart to the moon in 70 years. And we were only trying for the last 10!!!!

We didn't know what a germ was in 1880, we now transplant hearts daily. And for a substantial discount in Indonesia.

In 1880, news took six weeks to reach Seattle from New York; it now takes approximately 5 milliseconds.

The first research computers appeared just some 60 years ago.

All the above was accomplished in give or take 100 years.

Just maybe, in 1,000 years from now, surely we'll have reached an interim stage where we have a replacement for everything, including the brain which will have pre-installed everything: language (there'll only be one), reasoning ability, an enormous memory etc. etc.

Later, further out, we'll probably have said good-bye to flesh and blood, and its associated pain, stress and limited life span. Maybe we'll have switched to synthetic 'bodies' . . . . **"Mandroids."**

But what will we do in the following million years?

. . . .

I would certainly not be surprised if we discover that some of the UFOs, so much in the news today, are manned by 'droids' which have developed from alien societies from elsewhere in the universe. Now, freed of the frailties of their animal beginnings, they are ready and well equipped to travel the enormous distances through untold years in order to explore the universe . . . . **With whatever purpose in mind**.

Should we worry? Hopefully not, for what on earth could they want from us that they do not already enjoy? They already have all the amenities they could wish for. And, being droids, they don't eat meat . . . . hopefully.

And I still don't see any evidence of a god behind all this.

PART 5

WHERE DOES MAN FIT IN?

WHERE DOES MAN FIT IN?

In the preceding sections I have explored, to a modest extent, religion, the universe, evolution, some of the threats to man's future and a glimpse of what the early stages of that future might be. In each I have seen no evidence that any supernatural being or god has ever, or will ever, interfere with the known laws of the universe.

RELIGION

There is no denying that religions of all kinds have existed from the beginning of modern man and probably will exist for eons to come. As many before me have speculated, there seems to be a yearning in us to survive earthly death and an inability to accept that this life is all we get.

We gather together to provide some form of legitimacy to the need to believe there is more. In almost any situation, the more people involved the more credible becomes the 'group belief.' This would certainly explain why religions flourish in all parts of the world and among the most diverse races. I suspect that the 76% of polled US citizens affirming a belief in Christianity is really more like 10% actual believers and 66% sincere but 'me-too' groupies.

THE UNIVERSE

I examined the universe from its beginning at the 'Big Bang,' some 13.7 billion years ago, through to the present, and to what current knowledge and projections suggest will be its end, some 20 billion years from now.

I have even speculated a little on man's chances of surviving the dangerous events that might befall the earth.

The universe is an incredibly huge and wonderfully majestic place. Our knowledge of it has barely scratched the surface . . . but we will learn more. However, 'gaps' remain . . . we do not yet know what triggered the 'Big Bang' or what, if anything existed before. However, I see no reason to throw up our arms and invent a god to cut short our search for the truth.

What I found was that at all times, including times of extreme disorder, or even chaos, everything obeys the natural laws of physics . . . with no exceptions . . . . and no miracles.

EVOLUTION

I looked into the evolution of life on earth, from just after the earth condensed out of the remnants of the supernova explosion that preceded our solar system, through to the present.

Evolution has resulted in the incredible number of species that have lived on the earth since its formation, 95% or more of which are now extinct. I found no evidence of any outside direction or guidance in the evolution of life. Random mutations and the law of natural selection provide an incredibly elegant, simple and persuasive explanation. But again, there is another 'gap' . . . we have not yet learned what sparked that first small, but all-important jump from inanimate to living. There are recent claims that research scientists have accomplished this, I eagerly await more information.

Some conjecture that life may have started elsewhere in the universe and may have arrived on earth as part of a meteor. Maybe so, but it would not change my conclusions here.

GAPS

Yes, there are many events that have yet to be explained, and many even seem inexplicable, but that is no reason to say "God did it." I found no credible examples where the laws of the universe were set aside, overruled, or tinkered with.

When we speak of the 'laws of the universe,' it seems to imply that some entity must have 'made' those laws . . . far from it. The universe behaves in rational, logical ways and as man discovers these behaviors, we call them laws; simply because under the same conditions the results are always the same.

So, we still have major 'gaps' in our knowledge. However, in no way do I feel it necessary to assign a god to these 'gaps.' I fully expect that, one day, long after I have gone, these 'gaps' and many more that will surely appear, will be filled with scientific explanations. I would love to be around to learn them. But I've had my turn.

WHAT'S SO SPECIAL ABOUT US ANYWAY?

But, we ask, surely mankind is 'special' in some way; surely there has to be more after this life? Why? In all the religious cacophony about a future life one rarely hears any argument that any other animal will have an afterlife; we seem to be quite happy to let them disappear from existence. So why should we not just disappear too?

After all, we didn't exist before we were conceived, so is it really that unreasonable that we should just return to that non-existent condition after we die?

We are just one of millions of species that have evolved on this planet and, so far, have only occupied a microscopically narrow window in time. As pointed out in the time chart in Part 3 Evolution: **If the earth is one year old, then Jesus was born only 13.4 seconds ago and lived for just one fifth of a second.**

Putting aside our egos, there are only a few things that separate us from other species:

- We have a larger and more evolved brain, included in which is a sizeable memory. Even when selectively employed, our memory far exceeds the memory of other creatures.
- We are able to communicate freely and efficiently by voice, a quite recently evolved skill.
- We have an opposed thumb, which greatly enhances the use of tools and weapons. This feature was enormously important when we still had to compete with other animals for food and survival.

Nevertheless, our DNA is less than 1% different from chimpanzees, and I have read that an ear of corn has 50% more genes than a human being.

Had the asteroid of 65 million years ago missed the earth, we would probably not have made it passed snack food for the dinosaurs.

In other words, I do not believe that we have any differences to brag about that did not evolve by natural selection. I see no evidence of any supernaturally given (or imposed) differences.

MORAL AWARENESS

I have often heard it said that only man has a natural moral awareness. I do not agree; it seems to me that moral awareness is something that has evolved culturally, not biologically. A toddler, as it grows does not seem to have any innate sense of morality or moral awareness; far from it! A toddler is inherently selfish and, as any parent knows, it has a hard time learning to share. It has to be taught this sense of morality. Likewise my dogs had to be taught certain behavior traits. Once this is done, the head and ears go down when the dog is confronted with a misdeed. To me that is exactly the same as our much touted moral awareness.

With man's greater intellectual power has also come the ability to lie and deceive, another trait not found in other creatures, except perhaps in the search for food. In addition, man is the only animal that kills for pleasure. Who are we to speak of moral awareness?

I have witnessed 74 years of man's inhumanity to man (and to other species). I have watched many documentary films on the holocaust and torture and other nasty traits of the human that persist and even flourish today. In rare moments of despair, I am inclined to feel that the earth and its other occupants would be much better off after another 10-mile asteroid.

Yes, we are at a higher level of evolution than other living things, but all too often we abuse our 'superiority.' We often forget, or fail to accept, that all creatures have an equal right to be here. They too, were not consulted before being brought into this world, nor were they asked what form they would like to take.

Perhaps I exaggerate or take man's faults too seriously. Certainly, man's accomplishments have been incredible with many more, even greater ones, yet to come. And yes, there are indeed many pockets of great kindness and goodness in this world. I am not immune to feelings of great emotion when I see humans performing great feats of excellence in both sports and self-sacrifice for their fellow man. I often shed a tear or two when things go amiss for others, and lesser animals too.

It is my hope that mankind will find a way to improve its moral standards. It could, but there is little progress along those lines so far.

SMALL FISH IN A SMALL POND

We seem to have become so comfortable with our place at the top of the food chain and with our dominion over all other creatures, great and small, that we ignore the fact that we are but a miniscule minnow in a very small puddle, when compared to the rest of the universe.

How will we feel when a much higher (or more evolved) species arrives and treats us as we treat our less developed animal kingdom? I argued earlier in this work, that there are almost certainly species 'out there' far ahead of us.

So, unless we are threatened or harmed in any way by other creatures, perhaps a little more 'laissez faire' might be in order.

On the other hand, I am not a vegetarian and I often feel that I hold a double standard. Still, probably about 95% of all animals are somewhere in the food chain. And that is just what we are . . . another animal.

Clearly, at least to me, moral standards are not divine, but man-made and, it would seem, quite flexible.

AND FINALLY . . . .

This exercise has occupied only a few years of my life, but has had a profound effect on me. Even though I have never been a practicing

Christian, there has always been the thought that someday there will be some kind of reckoning. Not, perhaps, an interview with Saint Peter at the pearly gates, but at least maybe a coming face to face with oneself. This, in turn, produced in me a vague feeling that I should 'do something' about it and, at 78, soon.

As I have written these pages, I have come to believe that there will be no reckoning of any kind. There is no ship on the horizon to which I may send a belated 'I'm sorry, come and save me' message. It is not that it is too late . . . there never was any ship there at all! We are alone and, aside from social values, we answer only to ourselves. And this I'm ready to accept without hesitation, albeit with great emotional disappointment and sadness.

And so, to sum up:

I have not come across the slightest scrap of credible evidence to suggest that man is anything but an insignificant and temporary by-product of the evolution of the universe.

Despite this conclusion and belief, I am rewarded in that:

I no longer feel any guilt that I may not be conforming to the way the religions of the world would have me believe and behave.

I no longer trouble myself worrying over whether I'm building up enough credits to escape a bad afterlife or gain access to a good one, and I am certainly happier now that this, admittedly light, burden has gone.

I now feel much more tolerant of others, even though I have not 'walked a mile in their shoes.'

Believing, as I now do, that this life is all that we can expect, life has become much sweeter and more precious than it ever was before.

If I have a regret, it is that I could not find the time to think this out for myself earlier in my life.

I am now content to be an insignificant child of the universe, and look forward to whatever remains of my life, even if only a short time; for as Isaac Watts (1674—1748) noted:

Time, like an ever flowing stream,
Bears all its sons away.
They fly, forgotten, as a dream
Dies at the opening day.

www.ingramcontent.com/pod-product-compliance
Lightning Source LLC
Chambersburg PA
CBHW032027290526
45786CB00011B/788